The Dawn of a New Age

Reflections on Science and Human Affairs

Eugene Rabinowitch

CHICAGO AND LONDON
THE UNIVERSITY OF CHICAGO PRESS

Library of Congress Catalog Card Number: 63-20898

The University of Chicago Press, Chicago & London
The University of Toronto Press, Toronto 5, Canada

Published 1963. Composed and printed in the United States of America

In 1962, mankind came close as never before to the abyss of a nuclear war. Shall such paroxysms of power politics be permitted to occur again and again, until one of them ends in disaster? To prevent this calls for rebellion—not a rebellion in the streets, with guns or knives, but a rebellion of minds and consciences; not a rebellion of peoples oppressed by dictatorial rule, old or new, and not the rebellion of oppressed nations against alien rule, old or new, but the rebellion of mankind against a system which sets parts of it against each other.

This is a rebellion against rulers who, whether self-appointed or freely elected, consider it their right and duty to put the interests and ideals of the fraction of humanity, over which they rule, above the interests of mankind as a whole and above the moral duties all men owe to it—rulers who ask, from those they rule, absolute and ultimate loyalty to the national power and to the faith or ideology for which they stand.

A man can be asked to give everything he owns, his very life, for the protection of his family, his community, the nation to which he belongs, and for the faith or idea in which he believes. But a man cannot be asked to become a perpetrator or conniver in a millionfold murder of innocent men, women, and children, in the name of special power interests or a special faith or idea to which his group of mankind is committed, and to destroy in the name of his loyalty to this group the very basis of human existence on earth. Man owes his highest and overriding loyalty to mankind, not to any of its parts.

What should be the aim of the rebellion? To establish in all parts of the earth the rule of men who would consider their power over a part of humanity not—or not only—as a mandate to defend and advance the material and spiritual ideals of this part, but, above all, as a trust on behalf of mankind as a whole to contribute to its survival, to strengthen its viability, and to further its progress with all means at the disposal of the part of humanity which they represent. Men dedicated to this common aim above all others should arise in all parts of the world; and scientists should be the first to join this movement for survival and rebirth of mankind.

This exhortation followed the delivery of a formal paper, "The Whole before the Parts" (Part III, No. 21, in this book), at the London Conference on Science and World Affairs (COSWA-Pugwash) in September, 1962. It was later printed in the *Bulletin of the Atomic Scientists,* January, 1963.

Contents

PART I. LOOKING AHEAD

Introduction 3

1. Things To Come—Then (1939–41) 5

2. Atomic Weapons and the Korean War 12

3. How To Live in an Atomic World 16

4. After Missiles and Satellites, What? 28

5. First Things First 38

6. The Dawn of a New Decade 48

7. Lessons of Cuba 56

8. Things To Come—Now (1962) 70

Addendum: A Report to the Secretary of War 99

PART II. TAKING STOCK

Introduction 113

9. Two Years after Hiroshima 115

10. Europe in July, 1954 120

11. Ten Years That Changed the World 131

12. The First Year of Deterrence 146

13. New Year's Thoughts—1958 160

14. Hail and Farewell 171

15. New Year's Thoughts—1962 180

PART III. TALKING WITH WORLD SCIENTISTS

Introduction 189

16. International Co-operation of Scientists 193

17. *Russian and Soviet Science* 195
18. *Stop before Turning* 209
19. *Responsibilities of Scientists in the Atomic Age* 217
20. *Creation of a Suitable Climate for Disarmament* 227
21. *The Whole above the Parts* 245
Addendum: The Vienna Declaration 251

PART IV. HERETICAL THOUGHTS

22. *Science and Education for Peace* 261
23. *The Labors of Sisyphus* 266
24. *A Speech for the President* 269
25. *What Is a Security Risk?* 277
26. *Science and the Humanities in Education* 282
27. *Integral Science and Atomized Art* 298
28. *The Atomic Age Doctrine* 305
29. *Science, Scientists, and International Policy* 312
30. *Heroes of Our Time* 325
Index 329

PART I · LOOKING AHEAD

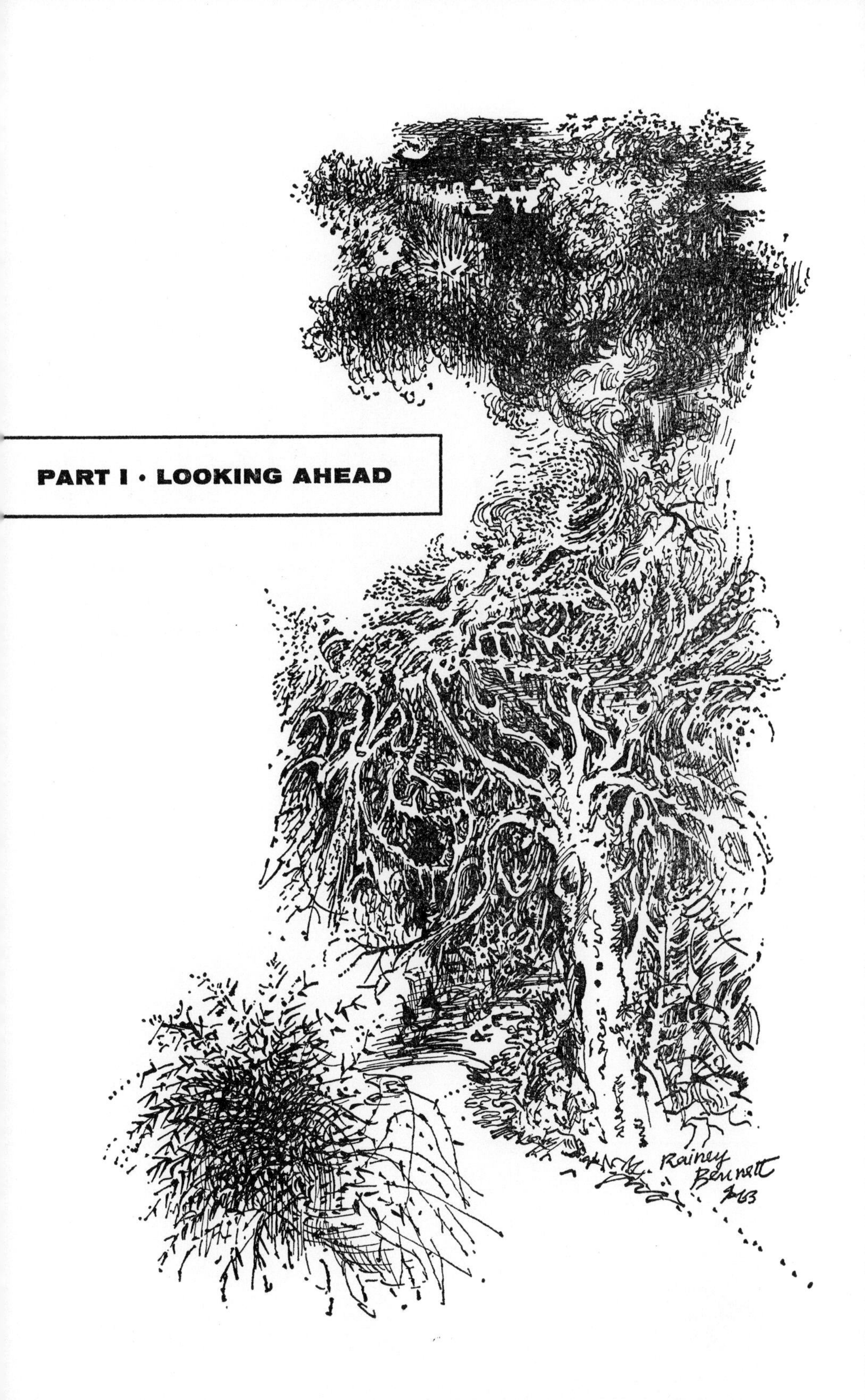

Introduction to Part I

The articles assembled in this part of the book have in common the intent of looking into the future, to foresee the shape of things to come. The first and the seventh of them, in particular, are frank attempts at prediction—predicting the course of the great fighting war in 1939 and of the great cold war in 1962.

In our time, scientific and technological factors are becoming increasingly important in human affairs. While current political developments remain determined largely by emotional attitudes of peoples and unpredictable personal decisions of their leaders, science and technology are changing before our eyes the very conditions of human existence, rearranging the stage on which the drama of history is played. They determine the development of mankind in its long-range trends, while the zigzags of day-to-day events remain irrational and unpredictable.

It has been said that the art of politics is the art of foreseeing the future; if this is true, success in this art now requires a growing understanding of the meaning and direction of scientific and technological change. Forecasting the future becomes less and less a matter of intuition and more and more a matter of objective evaluation of facts and dispassionate, unprejudiced estimation of masses and forces involved in political interplay.

It will be said that the following articles are far from fulfilling these requirements; that they do not deal with the scientific and technological aspects of political development and their quantitative consequences for human affairs; that they contain much guessing—forecasting by vague intuition—only camouflaged by some physical terminology. These criticisms are largely valid. The only thing claimed for these articles is that they represent an attempt to deal with political situations with as much objectivity, and as little wishful thinking, as scientific training has made it possible for the author and to use, as far as possible, a quantitative instead of the qualitative approach common to most political writing.

It is paradoxical, but true, that science, while it offers us an increasingly important tool for the prediction of the future, is itself as little predictable as are all creative human endeavors. In the article with which this section opens, the prediction of the course of the Second World War turned out to be rather successful because—like all preceding wars—this war was fought and won with technological means already at hand (or clearly in the offing) at the beginning of the conflict. It was a game played with a known set of chessmen and according to known rules. One event which could have fouled all predictions was the development of the atom bomb. However, it came too late in the war to affect its course or outcome; it only added a loud fanfare to the closing chords of the tragic symphony.

Science is one of the forces that can bring forth new factors, affecting, in an unpredictable way, the development and interplay of human societies. It is not the only such force; but in our time, which is that of a scientific revolution, new scientific insights and technological breakthroughs occur with unprecedented frequency and suddenness. This makes forecasting more difficult than ever—it is like predicting the outcome of a game in which new figures can appear on the board at any time.

1

Things To Come—Then (1939–41)

INTRODUCTION

The following piece was never published. I had written for newspapers as a college student from 1920 to 1925, but between 1926 and 1939, all my writing had been in the field of science. At the beginning of the war, on August 30, 1939, the idea came to me to write down some considerations about its likely course and outcome; but I did not then think of publishing these forecasts—only of giving myself the chance of doing what a scientist calls "checking theoretical predictions against experiment." I read this paper then to some of my friends, and their memory—by now probably dim—is the only support I can claim for the assertion that the piece is reprinted here as it was written then. The first section was written on the day of the German invasion of Poland, September 1, 1939; the second, in May, 1941, after the fall of France and the battle of England.

In 1947, I read this piece to some of my colleagues at the University of Illinois, and they urged me to publish it; but I felt that my own statement about the time of the writing would not be believed. Since then, my writing in the *Bulletin of the Atomic Scientists* has given me—I hope—some public reputation, and now, perhaps, my assertion that the piece is reprinted exactly as originally written will be accepted, at least by some of the readers.

An attempt to write a third part after the end of hostilities in Europe has remained in outline and is reprinted as such here.

It is easy to spot the errors—the invasion of Denmark and Norway was not anticipated, a German drive through the Balkans and Turkey into the Near East was predicted, and the cohesion (although not the tenacity) of Russian resistance was underestimated. The correctness of other forecasts and of the over-all expectations concerning the duration, course, and outcome of the war may appear self-evident now, a quarter of a century later. But rereading this piece strengthens my belief that viewing approaching world events with some of the detachment to which the study of science has accustomed one helps to avoid errors into which partisanship often leads the most astute political leaders and observers. I cannot resist the temptation of placing it at the head of this collection of essays.

WOODS HOLE, MASSACHUSETTS
September 1, 1939

Six years ago, a few days after Hitler came to power, I met (in Göttingen, where I was then working) a Nazi acquaintance of mine.

He objected to my distress, saying, "You should not look on all things only from the point of view of a Jew." I said that what distressed me was not only the probable fate of Jews under Hitler but the certainty of a new war. My friend laughed and said, "How can Germany make war? Look at the French army, at the British fleet." I answered: "That's why war will not come right away but only after Germany has had time to arm." And so we made a bet—I said war would come in seven years. Now, six months short of seven years have elapsed, and war has come. This makes me think: To what extent can one predict the future in politics? And it seems to me that one can do so with some chance of success if one does it without wishful thinking, particularly if the masses on the world scene have coagulated and the forces that drive them have become oriented. Then, the laws of mechanics can be applied to forecast the probable collisions and their results.

Applying this kind of forecasting to the present situation, it is clear that the impact of German mass, multiplied by the force behind it, will destroy Poland within a few, say four, weeks. It is also clear, if one avoids wishful thinking, that the U.S.S.R. has a pact with Germany which will probably permit it to march in and occupy the Baltic countries and eastern Poland, perhaps also Bessarabia.

After this division of the east, Hitler probably will again offer peace to England and France. But the forces now in action there will not permit English or French politicians to accept such an offer. Thus, it will end—probably in spring, 1940—with a German attack on the west. Because of the Maginot line and the mountains to the south of it, Germany will be forced to repeat the 1914 strategy of marching through the Low Countries. There is little doubt—comparing the quantities of motion (*mv*) that existed in the west in 1914 with those in existence now, and the absence of an eastern front—that the Germans this time will succeed in overrunning France.

Then, they probably will again offer peace to England, and will again fail, because of the persistence of British mass once set in motion. Germany has started the war without a navy; the fate of England therefore depends on whether the German air force alone will be sufficient to subjugate England or to make it untenable. The ratio of German to British air force cannot be judged at present, so no prediction is possible. But one thing is certain—even if Britain is overrun, the war will not end but will spread over the globe, involving in the subsequent years America, Russia, and Japan—the second center of aggression against the "haves" in the Far East. Considering the *mv*'s (manpower and industrial power times drive and leadership) available on both sides, this contest will certainly lead to a final defeat of Germany and Japan, but it may take a long time—perhaps five years if England stands and ten to fifteen years if it falls; and America will have to carry the brunt of it.

WATERTOWN, MASSACHUSETTS
May 5, 1941

Poland is divided, France overrun, England bombed but not defeated or invaded. My prediction in 1933, and again in 1939, was: *Germany will almost win the war, but it will be defeated at the end because it will overestimate its own force and underestimate that of its enemies—as it did in 1914–18—and strike before it is prepared to meet all eventualities.*

This prediction still stands. Germany struck before building a large naval force; she struck before achieving a degree of superiority in the air that would lay England defenseless at her feet. If Hitler had restrained himself before Prague and Danzig, he might have built up an air force much more formidable than the one he actually had at the outset of the war without arousing England into action. His error was in thinking that the strength he already possessed was sufficient to frighten the "effete" countries of the West into submission without a fight; he still clung to this hope when he offered "peace" after the conquest of Poland and again after the defeat of France. He might still entertain similar hopes in respect to America.

However, it is not in Hitler's power to choose the time of the conflict now. He cast the dice, and they are now rolling. The conflict will be decided on the basis of realities—masses and velocities—as they exist in the world *now,* not as they could be cleverly rearranged at another, later juncture. These realities are: Germany is, and will remain for a long time, an immensely superior *land* force. England and America are the superior *sea powers,* even counting Japan in. Land power cannot strike directly at sea power; nor can sea power strike directly at land power. *Air power* will be on the side of Germany for probably another year—until summer, 1942; the two sides will be about equal in fall of 1942; and the development will probably begin to work against Germany in 1943. Air power alone cannot bring *quick* decisions. Therefore, the *war will be long —it may easily last another five years.*

Even a conquest of England would not bring the end of the war. But *I do not expect England to be invaded;* I do not even expect an attempt to invade it, because the possibility of Germany gaining control of the air over England is now less likely than it was in 1940.

The Mediterranean, North Africa, and the Near East are likely to fall to Germany, since they are accessible to land power. An American intervention in the summer of 1940 could have prevented the French fleet and colonies from deserting the cause of the Allies; it would thus have insured the conquest of Italian North Africa and increased tremendously the chances of preventing the overflow of the German army into Africa. This chance has been missed. An American intervention *now,* with a direct attack on Dakar and an expeditionary force in North Africa, could still retrieve the situation; but obviously, no such action will be taken, and this chance will be lost too, thus prolonging the war for perhaps another two years.

It is probable that the German thrust through the Lybian desert will remain unsuccessful because of the difficulties of maintaining adequate supplies for a large mechanized force. But how can Turkey refuse the passage of German troops, and if she refuses, how can she expect to stop them? Therefore, the eruption of German hordes into the Near East appears inevitable. Egypt will be attacked from two sides, and the British fleet will have to be evacuated through Gibraltar or Suez. It is only to be hoped that it will be able to get away without too severe losses. This will probably happen about July 1, 1941.

Russia will do nothing about Germany's move through the Balkans because the only thing it could do would be to make war on Germany, and Stalin is not ready for a suicide. He might, however, move into Iran and attempt to seize the oil fields there, not because of a secret undertsanding with Hitler about the division of the Near East, but because of the advantage such a move may bring to his bargaining power. Hitler is not in the necessity of buying Stalin's co-operation any more, as he was two years ago, but he might still prefer to close his eyes on some looting Stalin may do for himself.

After the conquest of North Africa and the Balkans, the German military machine on land will have only one task left—the *subjugation of Russia.* Perhaps, before undertaking this, Hitler will once more attempt to throttle England into submission by using his newly acquired bases in West Africa and on the Iberian Peninsula. I think that this will be the period—in July or August, 1941—when Germany *will come into open conflict with the United States.* America will be unable to stay out once the communications in the southern Atlantic are endangered.

Once the United States is in it, the war will resolve itself into a gigantic contest of technology. Industrial production will be reduced under the aerial bombardment in England and in western Germany; the contest will thus be between the United States and the less accessible parts of Germany. All advantages in respect to materials, equipment, and labor will be with the United States, so that as early as 1942 the production of war material will probably be higher on the Allied than on the German side. However, Germany will have the advantage of being able to make *better use* of its production because of the absence of appropriate bases for Allied attack except the British Isles. This is when the loss of African bases will become bitterly felt. Probably, an American expeditionary corps will be sent to Africa to fight its way, together with British imperial forces, up the coasts of West and East Africa to reconquer North Africa. This operation may take the whole of 1942.

In the meantime, *Germany will attack the U.S.S.R., probably in spring, 1942:* first, because it needs occupation and victories for its two hundred fifty divisions; second, because it cannot afford to take the chance of its own industrial production's declining under the stress of a long war and thus losing the preponderant strength it had relative to that of Russia. Germany will have to dispose of Russia before the exhaustion stage is reached in the war with England and America.

The conquest of a large part of European Russia probably will be achieved in two to three months because of the predominance of German technology, leadership, and above all, a superior supply apparatus. After this time, the Germans will be at or in Leningrad, Moscow, Kharkov, perhaps even North Caucasus. If they adopt a wise policy in dealing with the occupied country, they may easily disorganize the whole Soviet state and bring the people solidly behind them. *But it is very unlikely that Germans will be wise.* More likely, they will treat the occupied country so brutally, starting to make space for Germans to move in, that they will thus provoke a guerrilla war, endangering their lines of communication and permitting the Russian army to acquire a new spirit of resistance and a measure of equality with the German army at the end of the extended German supply lines. In this way, the war in the east is likely to be dragged out, with the German army gradually being bled white in attempts to keep a vast, seething country subjugated and to maintain a continuous barrier against the Russian army over a three-thousand-mile-long front from Archangel to the Caspian.

The battle of the Atlantic will become so serious that *the United States will have to call its fleet from the Pacific,* giving Japan a go-ahead signal in the South Seas. The resistance of the Dutch East Indies and Singapore may delay the Japanese progress, but a spread of Japanese might over all of the Far East during this period is inevitable.

By 1943, North Africa will be in American hands and the reconquest of Asia Minor will begin. The cities of Germany and England will by that time be utterly devastated. In summer, 1943, Germany will probably begin to have the worst of the aerial attacks from bases in England and North Africa.

In 1944, Germany and the whole continent of Europe will suffer increasing terror, hunger, and finally, anarchy. The Anglo-American expeditionary force will then begin local landings—first in Italy, then in Norway, and finally, the main one in France. Germany's defeat will be sealed; but Hitler will retain a firm grip upon a die-hard fraction of the German people. His terror will assume unheard-of proportions, with mass executions of all who try to stop the hopeless war. England and America, utterly exhausted themselves—since they will also have the task of gradually driving Japan out of its vast occupation area—will find themselves, in 1945, confronted with a stupendous task of reorganizing a Europe (and Asia) fallen into anarchy, with the eastern half of Europe probably in the hands of Communists or other revolutionary groups, torn by fratricidal wars of a cruelty not known in modern history, fed on economic disorganization and hunger. Will they master the chaos or will the chaos engulf them?

CHICAGO
May, 1946

I. Crystallization of masses and forces now:
   1. Eastern bloc. Communism and Slavic drive *vs.* West. Russian imperialism. *Main frontier in future Europe: Lübeck to Trieste.*
   2. Western bloc: it may or may not solidify.
   3. American imperialism—newcomer on world stage.
   4. Nationalism—old and new—everywhere.

II. Even if these *mv*'s dissolve, the general situation will remain the same—national sovereignties, no organization able to prevent war. Who would have expected Italy to become the first aggressor after the last war?—*somebody* will become it now.

III. Technical change—capacity of civilization to destroy itself, dollar for dollar.

IV. Alternatives:
   1. War before Russia has bomb—outlook.
   2. War afterward—destruction of American cities.
   3. War with "somebody" later—if it were not for Russia,

we would have even less unity and international organization than we have now.

V. Only change can come from a new, large *mv*. World organization—lots of *m*, but little *v*.

2

Atomic Weapons and the Korean War

At this writing, the consequences of the Communist invasion of the Korean republic and of the United Nations-sanctioned intervention in defense of this republic cannot be foreseen. Hope is not yet lost that in the face of American determination to resist the Soviet leaders will decide to call the invasion off and order the Communist troops to retire behind the 38th parallel. They operate on a long-range time schedule and can afford losing face much better than countries with a vocal public opinion.

It is much more likely, however, that the Communists will do all in their power to prolong the Korean conflict—without spreading it—hoping to make this campaign as unprofitable, as disappointing, and as undermining for American prestige in the East as the Indo-china war proved to be for France.

Finally, it is not impossible—although not very probable—that, either in accordance with Soviet plans or despite them, the Korean conflict will not remain localized for long, but similar situations will erupt over all southeast Asia, if not the whole world.

Whatever the development in the next few months, one fact has already been demonstrated by the first weeks of Korean fighting: It is the utter uselessness of atomic weapons in the present stage of our power contest with the Soviet Union, a stage which is intermediate between the cold war of the past three years and the threatening all-out hot war. (It could, perhaps, be called "warm" war.) It is not unlikely that this period will witness a series of small wars waged by the Communist satellites or guerrilla armies in Asia and Europe, while the Soviet Union will remain formally at peace with everybody, proclaiming strict adherence to the principle of non-intervention.

A lone conservative member of the British Parliament has suggested that the way to finish the Korean War in a hurry is to drop an atomic bomb on the capital of North Korea. This suggestion is obviously absurd. Flattening the capital of North Korea, with or without killing tens of thousands of its inhabitants, would not destroy the fighting capacity of the Communist army; but it would

From *Bulletin of the Atomic Scientists,* July, 1950.

immeasurably strengthen their case in the eyes of the world and weaken the cause for which America is fighting. Having destroyed the cities of Communist-led North Korea, would the next step be to destroy Communist-occupied Seoul, the capital of the South Korean republic, whose freedom and independence we are defending? It is hardly necessary to add that the possession of hydrogen bombs, for which we are striving, would have been of no help at all in this predicament.

Essentially the same situation would confront us if war were to involve Communist China—as well it may—or any other satellite nation in the East or in the West. Defending Indochina against Chinese Communists or the Philippine republic against Communist-led rebels, or supporting Yugoslavia against an attack by Bulgaria, Rumania, or Hungary, or aiding Greece against a renewal of guerrilla attacks abetted by its Communist neighbors—in every one of these situations, we would not be able to resort to area bombardment of cities by atomic or other weapons as a means to enforce our political aims. Everywhere we would be facing the question of how to protect a country from subjugation, or liberate a country already subjugated by an expansionist, totalitarian political system, without decimating its people and destroying their wealth, their homes, and their beloved ancient cities and monuments.

To be successful in this kind of "war by proxy" with the Soviet Union, the United States needs considerable and adequately equipped mobile land forces, strategically located as closely as possible to the areas where an outbreak appears likely. This kind of preparedness may be more burdensome and repugnant to Americans (and more expensive) than preparations for "push-button" aerial annihilation of large cities and crowded industrial areas. It seems, however, that the first task now assumes a higher priority than the second one. The ultimate outcome of the power contest between the United States and the U.S.S.R. may be heavily, if not decisively, prejudiced by a series of "little wars," of the kind now fought in Korea, which may break out sporadically in the Far East, Middle East, and eastern Europe. The existence of atomic weapons is one of the reasons (if not the main reason) why the outbreak of a hot war between the two main protagonists on the world stage will long be delayed. To this extent, atomic weapons in our possession may be called (as was claimed by Churchill) guarantees of peace. But, while this precarious "peace" is preserved on the main front, we are suffering and may suffer more defeats on secondary fronts where our atomic weapons are of no use whatsoever. Unless and until we find a way to stop the world power contest with the Soviet Union, and to replace it by peaceful competition and collaboration, we will have to undertake without further delay a considerable expansion of our mobile land forces in various strategic areas, even if it means slowing down the all-out development of weapons of mass destruction, such as the projected hydrogen bomb, and if need be, with disregard of domestic inconvenience and a decline in the standard of living.

Another thing is important, if not decisive, for meeting the challenge of the "warm" war: an active propaganda (and an active policy, which is the best propaganda), intended to create in the countries whose freedom we profess to protect popular belief in our intentions and active support for our cause. It is a difficult task, particularly in Asian countries where liberal elements are non-existent, or weak, corrupt, and without organizational ability, or meek and lacking in fighting spirit. The memory, and continuance, of racial injustice and colonial subjugation by the Western powers, exploited by Communist propaganda, must be overcome if we want to see on our side, not only the rich city dwellers and feudal landlords or the majority of peaceful, submissive peasants, but also the active, awakened minority, imbued with the idea of national liberation and impatient to achieve complete racial equality.

Our present possession and continued supremacy in the field of atomic weapons, and the means of their delivery, is an important asset in that it makes the Soviet Union wary of steps which may lead to the outbreak of an open war. If, however, such a war should break out despite all deterrents, the question of using atomic weapons against Soviet (or Soviet-occupied) cities will confront America with a fateful decision, foreshadowed by our present dilemma in Korea. Of course, from a purely military point of view, the use of atomic weapons will be justified, since in contrast to undeveloped Korea, highly industrialized countries, such as the Soviet Union, Poland, or Czechoslovakia, or any part of Western Europe, present targets, the destruction of which promises great advantage in a war with the Soviet Union. And yet, then as now, we will be facing the same dilemma—whether to convert a fight against an aggressive and oppressive political system into a war of annihilation against the peoples dominated by this system (whether for thirty years, as in Russia, or for three years, as in Czechoslovakia, makes no fundamental difference) or to renounce atomic bombardment and seek victory by slower and more costly methods.

If we concentrate on fabrication of weapons of mass destruction and do not balance this development by the creation of a sufficiently large, well-supplied, and strategically distributed land force, we will run a double danger: We will be in danger of losing out in peripheral skirmishes with Soviet satellites, such as in the Korean War; and we will have deprived ourselves of freedom of decision in the event of an open Soviet aggression against nations of the Atlantic Pact. If we have nothing but atomic bombs with which to strike back, we will obviously be forced to use these weapons, even if the enemy does not do so first and even if our leaders have grave doubts about the political wisdom and moral justification of their use.

3

How To Live in an Atomic World

Not quite ten years ago, the first atomic-fission bomb killed seventy thousand people and destroyed two square miles of the city of Hiroshima. In order to use such a bomb to detonate a thermonuclear bomb (H-bomb), its power first had to be increased about thirtyfold, from an equivalent of fifteen thousand tons (fifteen kilotons) of TNT to five hundred thousand tons. The thermonuclear explosions themselves, engineered by Americans in the Pacific and by Russians in Siberia, developed a power of up to fifteen million tons (fifteen megatons) of TNT. Such a bomb could destroy two hundred square miles of a city and kill several million city dwellers. New York, London, Moscow, or Peking could be effectively detroyed by a single H-bomb.

After Hiroshima, public opinion was more alarmed by the radiation threat of the new weapon than by the more familiar menace of blast and heat. The fear at that time seemed exaggerated, since the fine radioactive dust engendered by the explosion of an A-bomb high in the air was carried up into the stratosphere and scattered there by winds until it ceased to be dangerous. But, in the second of the Bikini tests in 1946, radioactivity revealed itself as a serious menace in an *underwater* explosion; the target fleet was drenched with radioactive spray, making the surviving ships unfit for manning for months afterward.

In the thermonuclear tests, a much more ominous threat of radioactivity became apparent. These giant bombs produced a vastly increased amount of radioactive products. Furthermore, since they were exploded close to the ground, they pulverized enormous amounts of rock, converting it into a relatively heavy radioactive dust. This dust "fell out" downwind from the explosion site, injuring islanders and Navy personnel on the surrounding atolls and Japanese fishermen on a ship eighty miles away. The radio operator of this ship has since died, after months of lingering radiation sickness. If bombs of this kind were exploded close to the ground in a future war, their radioactive fall-out would endanger people within a circle (or an ellipse, depending on the wind) with a diameter of

From the *New Leader,* January, 1955; later included in *Alternatives to the H-Bomb* (Boston: Beacon Press, 1955).

fifty or one hundred miles—far beyond the reach of direct bomb damage.

The weak radioactivity which remains in the atmosphere or the ocean long after an atomic explosion cannot damage organisms directly exposed to it, but it increases slightly the frequency of mutations in their genes. This can cause no general concern so long as only individuals or small population groups are exposed; but in a future war, when thousands of A- and H-bombs might be exploded in various parts of the world, whole countries or whole continents are likely to be affected, and the genetic consequences of such mass exposure may well prove disastrous. A widespread increase in the rate of mutations, however slight, could throw out of gear the mechanism of evolution by which species are evolved and maintained in nature. It is a delayed, insidious damage, and it may take hundreds of generations for its fatal results to become apparent. Experimental study of such slow, cumulative effects is practically impossible, except on short-lived lower organisms, which are often less sensitive to radiation than higher animals. That is why geneticists cannot predict exactly what a certain radioactive contamination of air or water will do to human beings or to other higher animals, but the best qualified among them take a very somber view. In fact, some geneticists are worried about the genetic consequences of even the slight, transient increase in radioactivity which has been noted over wide areas after A- or H-bomb tests. Thus, within ten years from their discovery, nuclear weapons have developed into an immediate threat to hundreds of millions of human lives and to our whole material civilization and have become a potential menace to the biological future of the human race.

Military and civil defense against atomic warfare includes (1) the threat of retaliation, (2) early warning and interception of bomb carriers, (3) evacuation of cities in the path of the attack, and (4) reduction of vulnerability by peacetime dispersal of population and industry.

The threat of instantaneous and overwhelming retaliation is, at present, the main reliance of the West in the face of the growing atomic strength of the Soviet Union. When continental defense (a co-ordinated system of radar warning chains, interceptor planes, and batteries of rockets and guided missiles) was first proposed by scientists two or three years ago, the proposal was viewed askance by some in the Strategic Air Command, who saw in it an attempt to undermine the role of atomic air power as America's main answer to the Soviet threat. (This behind-the-scenes conflict played an important role in the campaign against J. Robert Oppenheimer.) Since then, however, the SAC has recognized that preservation of its own retaliatory capacity depends on an effective warning and interception system, and continental defense has become an integral part of America's defense planning.

Since no "atomic Maginot line" can be 100 per cent effective, a well-organized civil defense is a necessary part of a continental defense system. The long-obsolete idea of "sitting the attack out" in underground shelters, which until recently dominated all civil-defense planning in America, has been recognized as suicidal after the H-bomb tests and replaced by a plan of evacuation of the cities threatened by the attack (which is likely to mean all metropolitan areas in the United States). The success of such evacuation depends, in addition to thorough preparation (including provision of new, broad evacuation lanes), on sufficient warning time. The radar chains under development at present promise to provide two or three hours' warning of the approach of propeller-driven bombers and perhaps one or two hours' warning of the approach of supersonic jet bombers. Whether this is enough time for the evacuation of a city is a crucial question; to answer it specific planning and rehearsal of evacuation operations are needed in our major cities. The dislocation of national, economic, political, and social life in the wake of the evacuation of all big American cities (and the destruction of some of them) is easily imaginable; so is the havoc which could be wreaked by their repeated evacuation and reoccupation, which may become necessary in certain political and military situations short of actual atomic attack.

Much less painful—and more effective—than evacuation *after* the alarm would be peacetime dispersal of essential administrative, industrial, and transportation centers, until now unforgivably dismissed with a smug and unimaginative "impossible!" Conversion of highly vulnerable, congested metropolitan centers into loose clusters of satellite cities, separated by ten to twenty miles of open space, could make a crippling attack on America much more difficult. Of course such dispersal, apart from its costs, would mean disruption of many existing patterns of economic, political, and social life; but it would have an enormous advantage over the other defense programs in that, once completed, it would provide a measure of permanent security and lead to a relaxation of tension and fear which no increase in continental defense or retaliatory power could hope to achieve. Dispersal may also be the only adequate answer to the threat of intercontinental ballistic missiles with atomic warheads, which, experts say, are to make their appearance within a few years. Because of their high speed, such missiles would reduce the warning time to less than an hour, making interception much more difficult and evacuation after the alarm impossible.

Retaliatory atomic air power, continental defense, city-evacuation plans, and peacetime dispersal are parts of an integral defense program the first purpose of which is to discourage aggression by making its success uncertain and retaliation as certain as possible. In the face of the political obstinacy, ideological fanaticism, and growing striking power of the Soviet Union, there can be no question of relaxing in any one of these efforts; on the contrary, the development of continental and civil defense and of evacuation and dispersal programs, as yet in their very early stages, needs a great acceleration to catch up with the menace. This is bound to involve a large and permanent increase in direct and indirect defense spending, but no budgetary considerations or political and economic inertia should be permitted to delay or cripple these vital programs.

None of these defense measures singly, nor all of them taken together, can promise anything resembling the "military security" of the pre-atomic age. They can, however, establish and maintain for a number of years a highly unstable and precarious status quo in which the outbreak of war will be made unlikely by the terrible risk the aggressor takes in unleashing it—a status quo in which nations will continue to spend a large part of their national income on accumulating weapons of mass destruction in their arsenals, erecting more and more elaborate "electronic curtains" around their territories, and shaping their political and economic life so as to survive best in a blowup that may occur at any moment.

In this tense situation, the first, natural impulse is to attempt to reduce the danger by outlawing atomic and thermonuclear weapons. Soviet spokesmen have called—and many people of good will have echoed the call—for a commitment by all nations to abjure the use of nuclear weapons in war. However, the analogy with the restraint shown by all nations in the Second World War in the use of poison gas, often quoted by Soviet speakers, is inappropriate. Poison gas is an ineffective, cumbersome, and indecisive weapon; atomic weapons are immensely effective, easily transportable, and potentially decisive. The great powers are now hectically preparing for none other than atomic warfare and are revising their strategic and tactical plans correspondingly. We have this from Field Marshal Montgomery and General Gruenther of NATO; although the Soviet military leaders are not talking, there is no doubt that they, too, are preparing to place their first reliance on atomic weapons. Because of this paramount importance of atomic weapons for national strength, the only acceptable form of atomic disarmament would be one providing objectively adequate guarantees of the actual elimination of atomic weapons from national arsenals; and it is questionable whether such guarantees are now technically possible.

Shortly after the discovery of atomic energy in 1945, scientists were the first to call for its international control as a means of preventing the possession of atomic weapons by individual nations, and this idea was soon accepted in principle by all major nations, including the U.S.S.R. Nine years later, controlled atomic disarmament still remains the official aim of all nations, including the United States and the Soviet Union.

In the meantime, however, UN-sponsored negotiations on the practical implementation of the international control of atomic energy have been deadlocked since 1948. The Western governments, advised by their technical experts, maintained at that time that the only effective method of control would be a world-wide monopoly of all large-scale atomic industries, from smelting of uranium ore to the generation of electricity in atomic power stations. The U.S.S.R. refused to accept this principle, denouncing it as an attempt to dominate "socialist industry" by what they called an "international capitalist monopoly." The Soviet speakers asserted that effective control could be achieved by inspection of national and private atomic plants by an international agency, but they were unable to suggest a concrete system of international inspection which would be an adequate substitute for international ownership (or, at least, international management) of atomic industries.

Ever since the control negotiations bogged down, prominent individuals and organizations all over the world, from the pope to the prime minister of India, from the International Council of Churches to the Partisans of Peace, from the Parliamentary Union to the Red Cross, have not ceased to urge the leaders of the world to search for a compromise solution. Recent disarmament discussions in the United Nations have raised the hope that both sides in the controversy may be willing to budge from their previously adamant positions. But controlled atomic disarmament is a matter not only of political reasonableness but of technical feasibility, and this aspect of the problem seems somehow to have slipped out of the public eye in the current discussions.

The "Technical Feasibility Report" of September 27, 1946, prepared by the technical subcommittee of the UN Atomic Energy Commission with the participation of Russian scientists, was the only unanimously agreed-upon document in the long history of the control negotiations. At the time this report was prepared, the only thing to be controlled was the production of atomic explosives.

Stockpiles of these materials existed only in the United States and were too small to constitute a threat to peace, provided all further production could be put under proper control. The situation is different now.

According to authoritative statements, American stocks of atomic explosives are approaching a level at which further production will become, although by no means militarily valueless, at least not vital, since the materials on hand will be sufficient to destroy, perhaps several times over, every "worthwhile" target on the face of the earth. The Soviet stockpile probably is considerably smaller, but it, too, is much too large to be disregarded in the establishment of controls. Therefore, the technical feasibility of atomic disarmament now depends on a reliable inventory of existing stocks of fissionable materials. Considering the extremely small bulk of these materials, and the absence of penetrating radiations emanating from them, the only possibility of inventorying them is for the agents of the UN control body to be led to the stockpiles by national officials who know where they are located. Neither the West nor the U.S.S.R. can be expected to base its own atomic disarmament on the trust that the other side has not concealed a substantial part of its stockpile. In democratic countries, an international control organization could at least hope for some check of governmental compliance through information supplied by citizens; but in totalitarian police states reliance on such co-operation seems out of the question.

If this conclusion is true, then we must add to the appalling knowledge of the material and biological damage of a future atomic war the sad recognition that effectively controlled atomic disarmament has ceased to be possible and that all attempts to find a compromise solution leading to such disarmament are therefore bound to remain futile. Mankind will have to live, from now on, with unlimited and unchecked stockpiles of atomic and thermonuclear explosives piling up, first in America and the Soviet Union, then in Great Britain, and later in other countries as well. It seems that the only realistic form of "atomic disarmament" is, from now on, the cessation of further bomb tests (as proposed by David Cavers, of the Harvard Law School, and David Inglis, of Argonne National Laboratory, among others)—a simple standstill agreement, which could be effectively controlled by an international monitoring agency from neutral territory. But this could hardly produce more than a slight relaxation of international tension, and it is likely to prove unacceptable to the Soviet Union because of America's present lead in weapons development. It is hardly necessary to mention that President Eisenhower's "atomic pool" plan, however desirable it may be for various other reasons, is not in itself, and is not likely to lead to, a significant atomic disarmament.

One might suggest that an indirect way to atomic disarmament might be found through the abolition of the means of delivery of atomic explosives—bomber planes, guided missiles, atomic cannons. It is doubtful, however, that even a very elaborate control over such vehicles could have the same reassuring effect on the international situation that the elimination of atomic-weapon stockpiles themselves would have.

The attempt to stop the march of time at the threshold of the atomic age, and give mankind a chance to adjust its political habits to its changed technological habitat, has failed. The technological clock has inexorably advanced, and we are now entering the age of atomic plenty, from which atomic explosives cannot be banished by any ingenious political compromise. Having failed to slow down technical progress long enough for political realites to catch up with it, mankind is now confronted with the reverse task—to adjust its political ways to the inexorable facts of atomic-age technology.

It would be utterly unrealistic to suggest that this readjustment, the greatest in human history, can be completed overnight or, for that matter, in any now foreseeable future. Therefore, the first task of practical statesmanship is to preserve peace despite the continuation of the arms race, so as to gain more time for a permanent solution. It is a difficult assignment for national leadership: to channel increasing amounts of national energy and resources into a seemingly endless arms race, to insist on a demographic and economic shakeup of the nation in the name of military security, and yet to avoid fostering warlike attitudes at home and to preserve patience and reasonableness in relations with the outside world—in brief, to carry a big H-bomb and speak softly.

We must not forget that the paramount purpose of our armaments is to prevent war and would be defeated if we had to use them to win one. This does not mean that we should practice appeasement of the Munich type, which, we know from experience, only makes war more likely. Nor can the non-Communist nations relax their defense efforts to facilitate "coexistence" with the Communist world (if by this is meant not simply the absence of war and maintenance of diplomatic relations but mutual trust and a community of political interest). Never in past history has rivalry between two great world powers ended without one (or more) military showdown; and never was the probability of a peaceful resolution of a power contest smaller than now, when the rival power we face is a totalitarian empire whose ruthless leadership firmly believes in the historical inevitability of their conflict with the outside world and in their predestined victory and ultimate domination over all mankind.

Yet, if we fully realize the ultimate calamity for mankind of an atomic war, we will be patient, restrained, and always prepared for reasonable settlement. It is a tragic coincidence that the overwhelming necessity to put an end to war has arisen at a time when great peoples are suffering under a tyranny unprecedented in its brutality and all-pervasiveness, but we cannot change this fact. Only those whose rightful wrath and justified apprehension blinds them to the reality of atomic destruction can see in an atomic war a way to the liberation of nations now in bondage to Communist dictatorship. If "coexistence" means a renunciation of attempts to change forcibly a situation in which Communist dictatorships rule large parts of Europe and Asia, and a willingness to keep peace with these dictatorships as long as they keep peace with us, then coexistence is preferable to its only real alternative—war—and we must aim to prolong it and make the best of it.

The twin duties to arm as we have never armed before and to preserve peace as long as this is possible without compromising the power position of the non-Communist world are inescapable. But they should not be used as an excuse for leaving to future generations the task of evolving the answer to the great challenge which atomic energy poses to mankind—the establishment of permanent peace. This challenge has arisen in our generation, and we cannot plead being too preoccupied with Soviet imperialism to accept it. Unless we decide on our ultimate aims and use them as beacons to inspire ourselves and the world in the long, dreary years ahead, our efforts to avoid war by military preparedness and wise diplomacy will not bring us nearer real peace and will end as such efforts have ended in the past—merely in a postponement of the next war (which in our era means making it, when it comes, even more catastrophically destructive).

The question "Is there an alternative to the H-bomb?" has no positive answer if what is meant is "Is there an immediate military and political program which would make reliance on atomic and thermonuclear weapons unnecessary or permit their controlled abolition?" If, however, the question is whether an alternative exists to indefinite continuation of a world system in which peace is maintained by a balance of overwhelming destructive power in the hands of sovereign nations and by the fear which these weapons inspire, then the answer is Yes. This alternative is the establishment of a stable system of permanent peace on earth. The immediate consequence of choosing this alternative is the measurement of all our decisions and actions on the world scene by the new yardstick, "Does it bring us nearer to this ultimate aim?" rather than the traditionally supreme standard, "Does it advance America's national interests?"

Two—and only two—rational concepts of a permanently peaceful world are possible. One envisages a world of sovereign nations which have learned to live side by side in peace—an extension to the whole globe of the "security in disarmament" established in recent times between the United States and Canada, France and England, and Scandinavia and the Low Countries. Without formal acknowledgment, the policies of America and its Western allies can be said to be based on this hope. But the probability of this desirable state of international affairs being reached, if ever, without interruption by another war (and once having been reached, persisting forever) is extremely low. It presumes not only that nations now under aggressive totalitarian rule will follow a reasonably restrained policy until they finally evolve by internal transformation into "good neighbors" but also that in the future no nation anywhere in the world will come into the grip of an individual or a group who, for reasons of personal or national aggrandizement or hatred, or out of ideological or religious fanaticism, or by plain foolishness, will unleash a war or initiate a chain of events which will inevitably lead to war. The examples of Germany falling under the rule of Hitler, of Italy following the lead of Mussolini, or of the Arab countries permanently on the brink of war with Israel do not encourage such optimism, even if we could forget Stalin and his successors and Mao and his Asian allies. It must not be forgotten that, a few decades from now, atomic weapons are likely to find their way into the arsenals of all nations, great and small, enlightened or backward.

The other concept of permanent peace is the creation of a supranational world authority capable of enforcing peace. While it is patently impossible to create such an organization now, it is nevertheless possible and (be it submitted) imperative to dedicate the long-range world policies of America and its allies to this aim. In 1945, the unwillingness of America to relinquish its (largely illusory) sovereignty was as responsible for the introduction of the veto-provision into the United Nations Charter as was the distrustful isolationism of the Soviet Union. Now, ten years after the end of the war, we are in retreat from the point of closest approach to effective international community and to popular acceptance of our role as a part of it, which was reached in 1945. The European nations, shaken to their foundations by the war and psychologically ready at its end to accept merger in a larger organism, are regaining, with American help, political stability and economic prosperity in the framework of their old sovereignties. Political leadership acquires vested interest in these sovereignties, and peoples find again the satisfaction of "belonging" to them and the loyalty it engenders. This is the way of least effort, the old banks into which the waters return after the flood of war. We see it particularly clearly in the example of France choosing re-establishment of an independent German army in preference to a merger of European defense forces, even though the German chancellor himself warned that this decision is fraught with much greater danger of a rebirth of German militarism, and of a new war precipitated by it, than the rejected European Defense Community.

This trend in Europe (and in the rest of the non-Communist world) came about because America, at the zenith of its power and influence at the end of the war and in monopolistic possession of atomic weapons, did not realize to what extent these weapons have made a world system of multiple sovereignties obsolete and did not confront other nations with a blunt statement of this situation and an offer to discuss (and itself accept) an adequate alternative to this system. We are even less ready to turn away from the past now, after the postwar resurgence of nationalism in America and in the world. This is why an "alternative to the H-bomb" cannot be simply "picked up" but must first be created, and why the first step toward its creation—and thus toward the ending of an endless arms race and cold war—is for the American people and its leadership to commit themselves to a world community as their final aim. No Soviet imperialism, isolationism, or aggressiveness can prevent us from dedicating ourselves to this purpose, from acting in every situation in accordance with it, and from urging its acceptance by other nations. This is the only inspiring ultimate ideal with which we can oppose, in the forum of world opinion, the Communist ideal of a world federation of Soviet republics. The people everywhere know instinctively that what they need, above all, is peace; and however exuberant may be the nationalisms of some new (and some old) sovereign nations, the promise of permanent peace has a stronger appeal to all mankind than that of national independence and grandeur.

Man has become able to establish civilized communities only after recognizing the moral law binding each member of a community not to kill arbitrarily others belonging to the same community. This law did not extend, however, beyond the limits of the tribe or nation; it sanctioned and even glorified murder in intertribal or international war. For thousands of years, this partial restraint has been sufficient to permit the progress of mankind; some tribes or nations went down in defeat, but others prospered on the fruits of victory, and new civilizations grew beside or on the ruins of the old ones. Christianity, early in its expansion, adjusted itself to this limitation of the fundamental rule, "Thou shalt not kill," and so gained acceptance and recognition by nations and states. In the last reckoning, it is this deficiency of mankind's moral standards which confronts it now with the danger of decadence, perhaps even extinction. If man is to survive in the atomic age, he must learn that murder in the selfish interests of a sovereign state is as wrong as the murder of a fellow countryman in the selfish interests of an individual.

One may ask: Would this not mean the adoption of a philosophy of pacifism and lead to unilateral disarmament of the democratic nations in the face of a totalitarian power which is certain to see in it merely an invitation to conquest? The answer is that, on the ethical plane, we are confronted with as difficult—and as inescapable—a transition as on the political plane. The free nations have to maintain not only the maximum material defense establishment they are capable of building up but also the full devotion of those who have to man these defenses (and, in our age, of those who have to live in danger behind them). Yet, at the same time, they have to develop a new kind of moral foundation for this devotion. In the Korean War, for example, for the first time in history, we were not engaged in a war between sovereign nations but in an international police action. The fact that many Americans resented this designation as a depreciation of the valor and sacrifice of the American troops, who bore the brunt of the fighting, shows what is wrong with our present hierarchy of values.

Our education, our reading, and even our religion have accustomed us to think of fighting and dying for one's country as the highest expression of valor and devotion to community, whereas police action in support of law and order has a hard time being recognized as honorable even in many free countries. In contrast to the policeman, the soldier's and warrior's glory is the central theme of a large part of world literature, from the Bible and the *Iliad* to *War and Peace*. We will be capable, as a nation, of successful political action for a permanently peaceful world only in the measure in which we, as individuals, are prepared to support such action by accepting a new ideal and acknowledging a new set of political values.

This transition has already begun. Much of the literature born of the two world wars (except the state-controlled Soviet literature) expresses a deep disillusionment with the traditional values of national military glory. The Nuremberg trials were the first attempt to replace the age-old concept of war as a legitimate "continuation of national policies by other means" by that of war as an international crime. This attempt could not be fully convincing because even in the Western nations (not to speak of the Soviet Union) there has been no complete change of heart, no wholehearted subordination to a new supreme authority, no recognition of justice applicable to the victors as well as the vanquished, no readiness to remove national interests and national glory from their supreme position as motivations of national policies and to put the welfare and peace of all mankind in their place.

An "alternative to the H-bomb" cannot be found ready-made in new political strategy or new military tactics within the old framework of sovereign national states but must be evolved gradually, through day-to-day policies, from a change in the ethical standards of free peoples and their chosen leaders. That nations dominated by Communist dictatorships may—for the time being—be prevented by their rulers from even hearing about this change, not to speak of associating themselves with it, does not make it impossible for the free nations to initiate it; it only makes their responsibility to show the way more urgent and inescapable.

4

After Missiles and Satellites, What?

Ours is a schizophrenic age. We live on two different levels, in two incompatible worlds. In one of these worlds—no longer a dream world—all men, irrespective of their national allegiance, form a single community. Common dangers confront them: insufficiency (or threatened exhaustion) of natural resources, overpopulation and the hunger and sickness which they breed, misuse of science and technology. All men need permanent, stable peace. Their common interest calls for better understanding of nature and greater mastery of its forces. In the age of atomic bombs, jets, missiles, radio communications, and space travel, their fears cannot be banished, and their aspirations cannot be satisfied unless they all work together. An increasing number of people everywhere are becoming aware of this.

Since nations have been accustomed for centuries to living on an entirely different level, where group interests and international strife abide rather than co-operation and concern for the common weal, they can act on this basis only grudgingly and fitfully. In a venture onto the new level, sixty-four nations, urged on by their scientists, have come together in the International Geophysical Year. Recently, while carrying out the program of this collective undertaking, Russian scientists succeeded in launching artificial satellites, the first man-made objects to become cosmic bodies in their own right. The natural reaction of scientists everywhere is to rejoice in this feat and to admire and congratulate their Russian colleagues who have achieved it.

We have no doubt that the Russian scientists who launched the first satellites into space were inspired, primarily, by the urge to explore the universe around us. They sent into empty outer space instruments to measure cosmic rays, to follow the earth's magnetic field, to determine the frequency of meteors, to observe the sun's surface, in order to satisfy the human thirst for knowledge—not to help a group of men increase their power.

But the schizophrenic nature of our era makes it impossible to enjoy this triumph of human inventiveness without realizing that, on another level, joy and pride are changed to grim apprehension. On this level, where mankind is not a single community but an

From *Bulletin of the Atomic Scientists,* December, 1957.

agglomeration of quarreling fragments separated along national or ideological lines, the advancement of science within one segment becomes a threat to the others. The satellites appear, on this obsolescent but still frighteningly real level, grim announcements that one faction of mankind has increased its power to coerce the others. To all those opposing the political or economic ideology of this faction, and resisting its spread, it now appears (in the words of Trevor Gardner) that "destruction looks down their throats." Thus, the IGY, the shining effort of international co-operation, recedes into dimness, and instead there emerges the vision of a terrible danger to fragmented and feuding human society.

To scientists, it is clear that in our time the world community of interest is a living reality, the only possible path into the future, whereas the traditional world of international strife is only pseudo-real. It is a world in which the behavior of nations is determined by the shadows of once meaningful concepts—national sovereignty, martial glory, national security and defense, victory or defeat. Yet, scientists cannot ignore the continued power of these shades of the past to shape the attitudes of nations and the decisions of their leaders.

This is the tragic dualism of the role of science in the present world. Believing in the universality of scientific endeavor and the world-wide humanitarian purpose of science, scientists, at the same time, are loyal to the segments of mankind to which their birth, choice, or conviction have assigned them. They are obliged to provide the rulers of these fragments, elected or self-appointed, with arms which can be used for defense, deterrence, or aggression. This dichotomy will not end until mankind frees itself from the bondage of fragmented allegiance. Our Russian colleagues know this as well as we do. They will understand why American scientists, while congratulating them on their great achievement, are alarmed by its implications for the future.

The upheaval in the world and the storm on the American political scene caused by the Soviet satellites is a striking confirmation of the belief—on which the founding of the *Bulletin of the Atomic Scientists* was predicated twelve years ago—that science has acquired a decisive importance in human affairs. If our political leadership (and our public opinion!) had been more receptive to this idea, the American "Vanguard" project would not be bringing up the rear.

Did the beeps from the Soviet moon awaken the American people to the realities of the scientific age? The shock has been great and could be salutary; but the meaning of these events must be pondered far beyond their obvious, immediate lesson. This lesson was grasped immediately by almost all commentators, from Vice-President Nixon to the editorial writer of the smallest provincial newspaper. The few die-hard apologists of the "don't worry, boys, everything is under control" school—the Sherman Adamses, the Charlie Wilsons ("a neat scientific trick"), the Clarence Randalls ("a ridiculous bauble") —found little support even among the most faithful admirers of the conservative, paternalistic philosophy of a business-minded government. The failure of the creed of reducing federal spending, balancing the budget, and "letting private industry do the job" is too obvious. However, American public opinion forgets as easily as it is aroused. We cannot be sure that the effect will be lasting. The President has told the American people that they should "keep their chins up"—it sounded almost as if security were "just around the corner." With the passage of time (and perhaps with some evidence of internal troubles in the Soviet empire), bromides from high places may again revive complacency. How long will people strain their eyes to get a glimpse of a streaking point of light in a morning sky? Football games, race riots, and the doings of film stars are more exciting!

Probably the shock has been strong enough to give us a new, streamlined organization of the missile work. The appointment of Dr. James R. Killian is a step in this direction. But will America learn the deeper lesson of these events? Will it concede to science and learning the place on the national scale of values which alone can assure America a leading position in the scientific age? Will school curriculums be reformed? Will the financial and social status of the science teachers in school and college be improved? Will scientific achievement find the recognition to which only a few exceptionally dedicated scholars are indifferent and without which the brightest students are but little attracted to scientific careers? Will American leaders pay heed to what scientists have to say about the realities of the scientific age, about what science can do for (and to) ourselves and our descendants—if we mean to have descendants? Will films, radio, and TV cease to represent scientists as impractical dreamers, or as evil, heartless men, ready to experiment with everything without regard to human feelings? Will America understand that without rational, scientific humanism as one of its pillars, a free society will not survive?

Among the popular beliefs American scientists have been questioning all along, two are of particular relevance today. One is the smug belief in the superiority of American scientific and technological capacities; the second is the opinion that competitive private industry provides the best answer to all production challenges under any circumstances.

The best informed and most thoughtful American scientists have never underestimated Russian scientific capacities. One hears *ad nauseam* that "nobody expected the Soviet Union to acquire atomic bombs as quickly as it actually did," an alibi for an administration whose head, two years after the first Soviet atomic bomb test, still asserted that the Russians had no "real" atomic weapons. It is therefore appropriate to recall that in 1945 American atomic scientists predicted that it would take Soviet scientists four to five years to develop their own atomic bombs.[1] A few years later, at the time of purges in certain branches of Russian science, this editor warned: "It is wrong to think of contemporary Soviet science as being largely paralyzed by the ideological dictatorship of ignorant politicians. Some branches may be dead or stunted, but it is still a vigorous, growing tree. Limbs that have been cut off sprout new shoots; those prevented from growing in their natural direction grow around the obstacle."[2]

Particularly unfounded has been the American complacency about the Russian competition in rocketry. It is incorrect to speak of America's "losing its lead" in this field to the Russians. The fact is that America always has been behind Russia—tsarist or communist—in many areas of applied mechanics. We never made an effort to catch up. Russia has had an unbroken scholarly tradition in hydrodynamics, aerodynamics, and ballistics. It is ridiculous to seek the explanation for Soviet rocket success in the capture of German specialists. We got the most important of these specialists, and it was *our* military technology—and not that of the Russians—that needed this infusion of foreign blood to develop the art in this country. Even now, the most advanced rocket undertaking in America (the Redstone army team that produced the "Jupiter" and, according to recent newspaper stories, was about ready to produce a satellite launching rocket at the time that the satellite program was transferred to the navy) is in the hands of Wernher von Braun and thirty of his fellow German rocketeers, the men who created the terror weapons V-1 and V-2 in the Second World War. Traditional Russian strength in these fields of science has been as much a

[1] "A Report to the Secretary of War, June, 1945" (The Franck Report), pp. 99–109, in this book; Frederick Seitz and Hans Bethe, "How Close Is the Danger," in *One World or None* (New York: Whittlesey House, McGraw-Hill Book Co., Inc., 1946), pp. 42–46.

[2] Eugene Rabinowitch, "Science and Scientists in Russia," *Bulletin of the Atomic Scientists,* VIII (March, 1952), 74–78.

reason for their success as the Soviet government's early recognition of the military and political importance of rocket development. It must already have had high priority in the days when "Katyusha" rocket launchers helped turn the tide at Stalingrad.

Since 1945, when the Manhattan District of Army Engineers under General Leslie R. Groves "crashed through" in three years from the first nuclear chain reaction to the atomic bomb bursts at Alamogordo, Hiroshima, and Nagasaki, suggestions have been heard that this or that scientific aim could be reached in a hurry if a similar program were established. In general, scientists have not been enthusiastic about this idea. It is in the nature of science that truly great steps forward are seldom made in a hurry or even significantly accelerated by programming and organization. Even in our age of vast subsidized research projects, the most important advances remain the result of private enterprise of individual minds. This is not necessarily true, however, of practical applications of established scientific principles. Here, a large, well-organized, publicly sponsored effort, with no financial handicap, *can* bring about in a short time what private enterprise, hampered by competition and profit considerations, may be able to achieve only more slowly, if at all. If, in 1939, the atomic bomb development had been left to private industry, America would have had her first atomic bomb after, and not four years before, Russia.

The almost religious belief of the present administration in the superior efficiency of profit-motivated industry may be the main reason why, instead of catching up with the Russian lead in the rocket field after the war, we have fallen farther behind it. This interpretation may appear unreasonable—perhaps even subversive—to a business administration and a President carefully coached in the virtues of private enterprise. Despite the experience of the Manhattan project, they may be as unable to realize the limitations of their economic credo as Soviet leaders are unable to recognize the advantages of private enterprise in other areas. Dogmatic beliefs in the superiority of the socialist economy condemns the Soviet Union to backwardness in supplying its people with food, clothing, housing, furniture, and other consumers' goods. Equally surely, unquestioned belief in the greater efficiency of private industry handicaps American leadership in the realization of great technological aims whose practical achievement requires financial and logistic organization beyond the capacity of an individual firm.

Admittedly, there is no reason why America should have the lead in *all* technical progress. It can be argued, for example, that since

America has the least immediate stake in the development of industrial nuclear power, it does not need to strive for leadership in this field, leaving it to countries (such as Great Britain) whose need is truly pressing. Of course, even here, considerations of political prestige and American leadership may call for a national effort on a scale and with an urgency for which there is no justification in national economy. This has been the subject of a long and bitter controversy between the AEC, led by Mr. Lewis Strauss, and its congressional critics. Each side has strong arguments for its stand.

There can be no such doubts in the field of military missiles and rockets. America needs them more than any other country, and competition between different automobile makers or airplane firms will not produce them faster than they could be produced by a centrally directed government project. Similarly, if we believe that leadership in space travel is of crucial importance for America, future competition between Macy and Gimbel will not get Americans the first tickets to the moon, whereas a concerted national effort may.

However forcefully and urgently we try to repeat the Soviet rocket achievement, it seems likely that for several years to come our achievements in this field will remain less spectacular than those of our competitors. How calamitous can this situation become? The question has two aspects. One is the psychological effect on the standing of the United States, especially in the "uncommitted" part of the world. The Soviet satellite has done far more to enhance respect for the Soviet Union than did claims of having perfected an intercontinental ballistic missile. This may be because the existence of the satellite has been demonstrated for all to see, while that of the missile was merely announced in deliberately vague language. (Here, by the way, is a useful lesson on the relative merits of secrecy and openness in the cold war.) The sooner an American satellite joins the procession of Sputniks in the air, the less fateful will be the injury to the American reputation for advanced technology; but the former belief in our pre-eminence will never return.

From the point of view of actual balance of power—more specifically, of deterrent air-nuclear power—at the disposal of the two major antagonists, the satellite is of only indirect importance. What matters is the capacity for launching nuclear missiles at targets anywhere on the globe. The Soviet successes so far make it likely that this capacity will be acquired by the Soviet Union ahead of America. The crucial question is whether during the period of Soviet monopoly in transcontinental nuclear ballistic missiles, the American deterrent establishment, based on manned SAC planes, will retain its effectiveness. The race is therefore primarily between the American long-range missile program and the Soviet program of defense against manned bombers. A really dangerous imbalance will arise if, before American ballistic missiles are available in operational numbers, the Soviet military scientists have developed weapons able (in their belief) to deny the retaliatory threat of SAC bombers. More ominous than Khrushchev's proclamation of the Russian capacity to drop nuclear missiles on every point on earth is, therefore, his boast that America may as well scrap its bombers since the danger for such manned planes is (or soon will be) prohibitive. Here lies a real and critical danger; with the Soviet Union considerably ahead of the West in the development of *offensive* missiles, there is no reason to be complacent about their capacity to develop *defensive* weapons.

Admiration for the Soviet drive to develop rockets and satellites should not induce us to believe that Soviet science in general is superior to American science (not to mention Western science as a whole). Russia has had remarkable scientific successes in the past; in many areas of scientific endeavor an unbroken tradition of achievement goes back to the seventeenth century, to the days of Lomonosov. Under the Soviet regime, science has received the greatest encouragement and almost unlimited financial support. This has led to an immense quantitative expansion. Many fields, in which the seeds had been sown long before the revolution, rapidly blossomed. Older scientists, who had long toiled as individuals with little support were given the chance to create large schools. News areas (such as genetics and nuclear physics) were opened up by pioneers. Soon, however, Soviet science began to suffer under excesses of party orthodoxy; the young discipline of genetics was almost devastated by charlatans invoking the patronage of dialectical materialism. Stevan Dedijer, in a recent article in the *Bulletin of the Atomic Scientists,*[3] (which seems to have cost him his research job in Yugoslavia), suggested that only the well-established tradition of Russian science has permitted it to survive this harassment with relatively minor damage.

At first, the vast quantitative growth of Soviet science inevitably depressed the average quality of scientific work; but gradually this weakness was overcome, and sound achievements began to appear. Nevertheless, none of the greatest recent advances in pure science (physics, chemistry, or biology) originated in the Soviet Union. Only one Nobel prize has gone to Russia since the revolution—the chem-

[3] "Research and Freedom in Undeveloped Countries," XIII (September, 1957), 238–42.

istry prize awarded in 1956 to N. N. Semenov for his study of chemical chain reactions. During the same period, forty-seven Nobel prizes were awarded to America (including several that went to European scientists established in America after 1934). This disproportion is not a result of Western bias or ignorance of Russian achievements; rather, it is a fair reflection of relative contributions to the progress of fundamental science. It shows that the gap between quantitative expansion and qualitative perfection has not been quite closed in Russia in the forty years since the revolution. It may be argued that Soviet education, however well it prepares a future scientist or engineer for competent work in his field in research or industry, is not very conducive to awakening his mind for the unorthodox, fearless, critical thinking which is required for really epochal discoveries (such as those of evolution, relativity, or quantum theory).

The obvious first lesson of the Soviet rocket success is that America must put a maximum effort into its own rocket development, unhampered by budgetary considerations, service rivalries, or hesitations about using "socialist" methods instead of relying on private industry. This lesson, we hope, will stick. The second, medium-range, lesson is that America should give more scope to science and scientists, in education as well as in social and political life. This lesson has yet to be learned. This is, however, not all. It would be a tragedy if, in the face of the challenge presented by Soviet science and technology, and the necessity to match it in every important field, we postpone indefinitely all thinking beyond the meeting of this challenge. America made a similar mistake when it refused, during the war in Europe, to think beyond the day of victory. In the same nearsighted way, American leaders refused, before Hiroshima and Nagasaki, to listen to scientists concerned with the long-range consequences of the use of atomic weapons in Japan. The time to think about our ultimate hopes and aims, beyond the winning of the arms race, is now—in fact, it is long overdue.

When the difficulties of reaching an agreement with the Soviet Union in the United Nations on international control of atomic energy became apparent in 1947, some American scientists made an attempt to reach their Russian colleagues to discuss this problem with them. Through official Soviet channels, they were informed that Soviet scientists were "too busy to talk about politics." Recently, when this unknown episode in history was quoted to a prominent Russian scientist visiting this country (who complained that the atomic arms race should have been stopped at the very beginning), he replied, after a silence, "Yes, we were busy then—busy catching up with America." Because in the eyes of the Soviet leaders—and perhaps in those of the Soviet scientists as well—catching up with America appeared more important and urgent than exploring ways to prevent the atomic arms race from getting under way, the opportunity to do so was lost. The Russians did not want to "negotiate from weakness." Therefore, serious negotiations about the abolition of atomic weapons were delayed until Russia had its own atomic weapons, at which time it was too late for effective control.

The American reaction to the revelation of the Soviet lead in missile developments is similar. The matter of first priority is now to catch up with the Russians; it seems almost improper to talk about anything else: about the terrifying implications of a race in space weapons, whoever may be ahead. American scientists are "too busy" to think about these implications. By the time both we and the Russians possess operational intercontinental missiles, with hidden, well-protected bases for launching, their abolition will become technically impossible and militarily unthinkable, since the strategy of both sides will be built around them.

In 1946 an American physicist, Louis Ridenour, published in *Fortune* a whimsical but prophetic piece.[4] He saw the earth surrounded by a swarm of atom bomb-carrying satellites owned by various nations. By pressing the proper button, one of these deadly moons could be made to drop on a target anywhere in the world. In well-protected underground operation centers, the military of every nation was keeping day and night watch, with their eyes glued to radar screens and their fingers on the fateful switches. An earthquake in California, a misinterpretation of the blip it produced on the radar screen, a snap decision of an overwrought officer—and the rain of atomic satellites began to fall upon the cities of the world, as one nation after another was drawn into the deadly chain of massive retaliation. Ten years later, this vision is beginning to look ominously realistic. In a year or two, dozens of satellites will be flashing through the skies, some harmless (or even useful, like those planned for the IGY), some observing and photographing every corner of the world in preparation for a missile war. A few years more and Ridenour's fantasy of remotely controlled nuclear bomb carriers may become reality. Perhaps, manned space stations from which atomic rockets could be discharged to reach their targets in a matter of minutes, if not seconds, will be sailing around the earth.

A lawyer has recently asked for new laws to define the limits of

[4] "Pilot Lights of the Apocalypse," XXXIII (January, 1946), 116.

national sovereignty in the sky. How far into the universe does the U.S.S.R. or the United States, Thailand, or Monaco extend? Ten miles, a hundred miles, a million miles? A physicist answered, half jokingly, that perhaps national sovereignty should extend indefinitely into space, dividing the universe into conical segments with the different nations of the earth at their apexes. As the earth rotates, these cones will sweep the sky. For a brief interval, the sovereignty of the United States will encompass the planet Venus or the star Sirius; then it will be replaced by that of Egypt or Venezuela.

Extending national sovereignty into interplanetary space would of course produce continuous complaints of violations and invocation of force to protect it. The physicist suggested that the main sport of nations in the future may be shooting down satellites and missiles violating their cosmic space. It would be a merry contest of science and art, and the United Nations could keep impartial score!

This picture of the future may be whimsical nonsense; but in straining our energy and inventiveness to draw even with the Soviet Union in the possession of long-range missiles and the capacity to place them on targets anywhere in the world, we may too easily forget to think what the world will be like after we have successfully achieved this purpose. It will be a world constantly tottering on the brink of disaster, with survival depending on the permanent sanity and restraint not only of the leaders of sovereign nations but—as in the vision of Ridenour—also of various subordinates in key posts, who will have no time to refer the fateful decision to the capitals.

While we have no choice now but to try to maintain an equality in the contest for deterrent power, we should not delay thinking of how we can ever escape from the nightmare of nuclear space deterrence. We must not delay convincing first ourselves, then our leaders and our people, of the absolute necessity for ending the political fragmentation of mankind. Mankind must emerge from the infantile age of quarreling and fighting into the mature period of recognition of its community of interests. We must make it clear to ourselves and the world that America intends in the future to live in the world of the future and to break away from the fetters of the past. It is time to ask, and to keep asking, for an answer to the question: After missiles and satellites, what?

5

First Things First

Wars are fought with weapons. Since it seems that no nation can afford a war—modern weapons having made it too destructive—why not get rid of weapons? They have become terrifically expensive, so that, in order to hold its own in the arms race, a nation must spend a large part of its income and occupy a large proportion of its population with preparations for war. What could be more reasonable than to stop wasting funds and resources getting ready to fight a war nobody wants?

Disarmament seems to practically everybody the first order of international business. This is one point on which Khrushchev and Eisenhower, Adenauer, Macmillan, and Nehru sincerely agree. The joint United States-Soviet communiqué (September 27, 1959), issued after Khrushchev's visit, stated, "The Chairman of the Council of Ministers of the U.S.S.R. and the President of the United States agreed that the question of general disarmament is the most important one facing the world today. Both governments will make every effort to achieve a constructive solution of this problem." Why then has no progress been achieved in disarmament negotiations, which have been going on for years?

In his recent speech to the United Nations, proposing total disarmament in four years, Khrushchev said that if disarmament were achieved the Soviet Union would be willing to pool its efforts with those of the West in directing the released funds and technological capacities to the advancement of the underdeveloped parts of the world. Similar thoughts have often been expressed in the West: If only we could reduce our weapons budget, how much money we could apply to constructive efforts in all parts of the world! This is obviously true—but unfortunately irrelevant to the situation; and often, it is largely a pretext for not doing more for constructive purposes right now.

The arms race and the difficulty of effective control of any disarmament pact have the same—obvious and well-known—reason: it is the distrust between nations and societies. This distrust is in turn due to the justified apprehension, if not certainty, that the

From *Bulletin of the Atomic Scientists,* November, 1959. This article is reprinted with the permission of Basic Books, Inc., which has included it in a recent anthology, *Seven Minutes to Midnight.*

political aims of the various states and societies are diametrically opposite; in fact, that the main interest of one state or society is to weaken, if not to destroy, the other.

This kind of "national aim" has been traditional in all human history. Antique and modern states and empires have been built up by pursuing such aims and have been destroyed when they failed in the power contest. Differences in religion or in political philosophy, such as existed between Moslems and Christians during the Turkish and Arab invasion of Europe, or between the Protestants and Catholics in the Thirty Years' War, or such as exist now between the Communist and the libertarian societies, prolong and exacerbate such power conflicts, but these conflicts existed also—and assumed the most violent form—between Athens and Sparta, Rome and Carthage, England and France, western and central Europe, not to mention the free-for-all of the Napoleonic era. The conviction still survives in all parts of the world that power conflicts are the natural and permanent content of history.

The Soviet Union recently published transcripts of conversations between Stalin and De Gaulle in December, 1944. The head of Soviet communism and the leader of French nationalism had no trouble agreeing that Russia and France had one common interest—that of keeping Germany permanently at bay by combining their military power! Under the same leadership, France is now allied with Germany and opposing the expansionism of Soviet Russia. The roles have changed, but the play is still the same: the endless conflict of power interests.

As long as different nations pursue their national interests as their supreme aims and only occasionally pool their efforts in the pursuit of common interests (and this usually only when these common interests require opposing a third nation, as in the NATO pact), disarmament will not come. If a man knows, or reasonably believes, that his neighbor's purpose is to take away his property (if not his life), he will arm himself with the most effective weapons he can acquire. Of course, evident existence of effective police protection could persuade him to give up these weapons; but to hope for the establishment of an international police force, able to enforce proper behavior on nations as powerful as the Soviet Union or the United States, is to misunderstand altogether the state of affairs in the world. A police force can be established among individuals in a society because the great majority of them have a common interest in preventing violence and in protecting peaceful economic intercourse between them, and this majority is much stronger than the few individuals seeking to enrich themselves by violence. An international police will not be possible as long as the society of nations is dominated by the very nations against whom the police force will have to be employed.

Community of interests is the key; disarmament and a world police force can come only after common interests have been widely recognized as exceeding in importance the divergent interests of the different nations. We trust those whose interests we know—or believe—are largely common with ours. Despite all the mutual accusations, particularly at election times, the Republicans trust the Democrats and the Democrats trust the Republicans, the Conservatives trust the Labourites and the Labourites trust the Conservatives, because they know that the areas of their common interest are more fundamental than the areas of their conflict. The Communists did not trust the anti-Communists, and vice versa, during the civil war in Russia because both sides were convinced that there were hardly any common interests between them. This is why Lloyd George's plan to have both sides in the Russian civil war meet at Prinkipo to discuss peace was so pathetically absurd.

Common interests are established through co-operation. Or perhaps it would be more correct to say that while common interests always exist, people become aware of them only by the practice of co-operation. If nations postpone co-operation until after disarmament, disarmament will never become possible, however strongly and honestly everybody wants it. It has become a truism in the West, to which the East at least pays lip service, that disarmament is impossible without controls; but without mutual trust, controls cannot be effective. The East now suspects extensive controls as an espionage tool, and the West is obsessed with the danger of evasion from even the most extensive controls imaginable. Cessation of nuclear tests was originally considered the easiest thing to control; except for the extreme case of low-energy underground tests, all nuclear tests can be discovered by remote monitoring stations outside the country in which the test takes place. Yet, the possibility of this one kind of evasion has kept the international negotiations in Geneva frustrated for a year. In this case, the limits of uncertainty are so narrow, and the importance of possible evasions so questionable, that an agreement is likely to be reached in the end; but controls of really significant disarmament steps will be much more difficult, the acceptance of controls short of 100 per cent effectiveness much less likely, and the requirements of an evasion-proof control much more difficult.

The area of disarmament in which progress is least likely to be achieved, unless a very large degree of mutual trust is first established, is the abolition of arms which now assure each side in the world conflict the capability for a "second strike"—retaliation if it should be attacked first. These are, above all, the strategic air force and rockets with nuclear warheads. Such weapons could be abolished only under the most reliable controls, impossible without extensive mutual trust. It is therefore futile to think of beginning disarmament with the abolition of "weapons of mass destruction," as the Soviet Union has long insisted. Significantly, Khrushchev's speech to the United Nations indicated recognition that proposals to begin disarmament with "abolishing nuclear weapons" are impractical. In this speech, for the first time, the elimination of nuclear weapons (and of their carriers) was relegated to a later stage—the third—in the proposed sequence. The speech also contained a hint at the understanding of the present importance of foreign bases for the American retaliatory capacity by postponing the relinquishing of these bases to the second stage of the disarmament program.

Perhaps this increased realism of the Soviet approach to the disarmament problem may have something to do with the discussions of these topics between Russian and American scientists at the several Pugwash meetings. This, of course, is difficult to prove; but, be that as it may, the change is worth noting; and it is strange that apparently none of the official or unofficial commentators has noticed it.

There are areas of armaments, in which, at the present time, negotiations may be more promising. In the first place, all types of armaments which have lost or are losing their decisive importance could be considered. This means not only cavalry or battleships but may mean in time manned combat aircraft, which is being put out of business by the development of missiles. (This is strikingly demonstrated by the recent cancellation by the United States of a billion-dollar order for a new type of jet fighter and by the apparent lack of new combat plane developments in the Soviet Union.) At some future time, the same may even apply to submarines or other weapons of attrition in a long war—although the United States is not likely soon to renounce the possession of nuclear submarines, with indefinite cruising capacity and ability to launch intermediate-range rockets with nuclear warheads, since such submarines are important deterrent weapons.

Perhaps the most crucial aspect of the arms race, in which the interests of the Soviet Union and of the United States coincide—although the Soviet Union may not as yet be convinced of it—is the avoidance of what has recently been named "pre-emptive war" (to avoid the odious term "preventive war"). The danger of a war of this type arises from the present vulnerability of retaliatory weapons. The bombers on both the home and the advanced bases of the American Strategic Air Command, which are at present the mainstay of American deterrent strategy, are highly vulnerable; and many, if not most, of them could be put out of action by a sudden attack, particularly with long-range missiles. This is even more true of bases from which long-range, liquid fuel rockets of the "Atlas" and "Titan" type could be launched. In the face of this kind of vulnerability, military planners on both sides cannot but ask themselves: How can we prevent such a disaster to our retaliatory forces? A preventive attack may appear to them as the logical solution; but neither the government of the United States or the Soviet Union is likely to accept such advice because the risks of even the most successful preventive attack are too great to resort to this dangerous, violent solution in preference to a policy of non-military advancement (which at present must appear particularly promising to the Soviet Union). We Americans are convinced that unleashing a preventive war is impossible for a democratic state; the Soviet speakers assert that the same is true of a socialist state. But, independent of such considerations of political morality, the risk of launching a preventive attack is likely to be—and remain—too high for even the most ruthless leadership to resort to it in the midst of a "cold peace."

One cannot be equally certain about situations of crisis. In the past, in acute situations such as the Quemoy attack, the Suez expedition, or the Lebanon landing, the United States did practice "brinkmanship" by sending the fleet out to sea and alerting the strategic air force, so as to leave Peiping or Moscow in no doubt that we were actually ready to fight. There was not much risk involved in these actions since, in the absence of operational long-range missiles, the Soviet Union could only respond to them by similar preparations of its own—unless it actually wanted to unleash a war, say, by an attack on the United States fleet covering the landing in Beirut.

The Strategic Air Command, however, has long entertained the idea that it is not going to wait for enemy missiles or bombers to rain destruction on its bases. Rather, it would like to strike first, when it becomes clear that the other side's preparations indicate that an attack is imminent. This is the strategy of "pre-emptive war." It is known that the book by the SAC commander, General Power, the publication of which was recently prevented by the secretary of defense, contained an exposition of this military doctrine. There is no reason to assume that the President and the National Security Council have accepted this doctrine; but there is no doubt that the Strategic Air Command is dedicated to it and that their reasoning is convincing enough to make its adoption in some future contingency feasible.

Only when the United States acquires an arsenal of solid fuel rockets which could be fired rapidly from launching pads deep underground or from movable platforms at sea or on land will the arguments for pre-emptive attack lose their persuasiveness; but this will require several years.

One needs no secret sources of information to be sure that the same arguments for pre-emptive war occur also to military planners on the Soviet side. In fact, several recent publications confirm this supposition. One may argue that the Soviet military leadership has less need to resort to pre-emptive attack because it is much better informed about the development and location of our armaments than we are about theirs. However, the lesser risk of Soviet misinformation may be balanced by the greater restraints imposed on American leadership through the complex mechanisms of decision-making in the United States government, since in the Soviet Union, all the military may need to do is to convince the top political leadership that a pre-emptive attack must be launched.

In any case, it seems to me that the greatest danger of war in the next few years lies right there. A political crisis may arise—and this is likely to happen not once but repeatedly, if the experience of recent years can be extrapolated to the immediate future—and the situation may call for an effort to persuade the other side that we are not going to yield to pressure, and this may mean starting preparations for an armed conflict. The military on one or the other side (or on both sides) may then come to their political bosses and tell them: "According to our information, 'they,' on the other side of the ocean, are getting set for an attack. The only way we can guarantee that this attack will not be terribly destructive, and especially that it will not cripple our retaliatory power, is for us to strike first. The time for a pre-emptive strike is now!"

If this estimate of the danger is correct, there are two ways of reducing it. One lies in the political area. Both sides must recognize that the days when the game of brinkmanship could be played with some confidence—because war could be called off at any moment before or even after its start (as the Suez war demonstrated)—are over. Both sides must refrain from pressing a controversial issue until war appears imminent. Perhaps the tenor of Khrushchev's conversations in the United States (and in China) is evidence that this point is being recognized in Moscow; but I wouldn't be sure that this recognition is deep enough, on both sides, to prevent controversies, such as that about Berlin, from flaring up dangerously in future.

The second answer to the danger of pre-emptive war is military. Each side must recognize that it is in its own interest to prevent misunderstanding of its intentions by its opponent. This means that secrecy in arms development and military preparations should be considered not only from the point of view of the military advantage it confers on those who keep the secret but also from the point of view of the risk it entails that a potential enemy may not be able to distinguish preparations aimed at securing a capacity for "second strike" from preparations for a "first strike."

Whether such a distinction is at all possible—whether the type of armaments involved and the manner of their deployment are different enough to make it feasible—is far from clear to me, but I believe strongly that this problem calls for international study by experts and, first of all, for mutual understanding by the leaders of the great powers that, in this area, their interests are common. At the time of the abortive Geneva conference on the prevention of surprise attack, the Soviet negotiators showed very little understanding of this point of view. They wanted to speak about elimination of political conflicts, which they thought might lead to an outbreak of war, while the Americans wanted to talk about control of the kinds of weapons suitable for sudden attack.

The situation has changed since. The emergence of the concept of pre-emptive attack, in the American as well as in the Soviet press, shows clearly what kind of danger both nations are confronting. Nevertheless, there is as yet little evidence that the Soviet leaders have recognized that secrecy can be not only an asset but also a source of danger. Khrushchev in his conversations in America said that secrecy is a Soviet advantage, which it will not relinquish until well advanced on the disarmament trail; and even in the meeting of Western scientists with their Soviet colleagues at Baden this summer, one of the leaders of the Soviet group said very categorically that secrecy has nothing to do with the danger of war, that no nation has any business to know the weapons secrets of the other, and that secrecy of weapons development has always existed and will always remain. (He did not talk about espionage, but if this topic were mentioned, he would probably say that it, too, has always existed and will always remain.)

The realistically estimated possibilities of disarmament are thus limited. They extend to weapons whose importance is either minor or decreasing and whose abolition could be therefore accepted even with less than 100 per cent secure controls. They extend to nuclear weapons tests because the need for continued tests is minor, controls are relatively easy, and small possibilities of evasion can be tolerated.

In due time, they may extend to weapons which are important only for "old-fashioned," prolonged war, such as manned planes, tanks, and submarines. The chances for disarmament are nonexistent—for the time being—as far as nuclear weapons and missiles are concerned, although these weapons, the most expensive and the most destructive man has ever devised, are primarily responsible for the present universal clamor for disarmament. Perhaps there is some hope for an agreement on the specific nature and deployment of these weapons which could provide assurance that they are not intended for a "first strike" but are adapted only to the task of a retaliatory strike—if this distinction is technically feasible.

I do not want to appear to be "against disarmament"; this would seem as wicked as being against happiness and prosperity. I believe that the search for mutually acceptable disarmament steps must be continued with all possible good will and open-mindedness and that it deserves wide popular support. However, I also believe that to consider disarmament as the first and main item on the international agenda, and to expect great successes in this area, would be to deceive ourselves and the world. I believe that the first and main order of international business is to reduce the conflict of interests between the great powers and to replace it by a growing partnership.

As long as the main interests of the two power camps are opposite, and the possibility of an open power conflict between them remains real, most if not all of the local territorial conflicts between them are insoluble because—as Leo Szilard has often emphasized—every solution is bound to involve a change in the relative power of the two competing sides. (The Austrian settlement is an exception that does not invalidate the rule.) The best that can be done at present is to agree to postpone these problems until the context of the world situation is changed, either through growth of a universal community of interests or—probably earlier—through a change in the military aspects of the power conflict, which may reduce the importance of maintaining certain alliances, possessing certain advance bases, or having space for maneuvering outside one's home territory (as the Soviet Union now has through its domination of Eastern Europe).

We must watch for these changes and not cling stubbornly to security concepts which may become obsolete; and this, unfortunately, may be more difficult for a democratic country, where learning and unlearning must be done by large numbers of people, than for a totalitarian system, where only the views of a small number of people, although particularly stubborn and opinionated people, must be changed. But, the most important thing we can and must do is to pursue, without hesitation and delay, stubbornly and imaginatively, the aim of enlarging the areas of common interest and co-operation.

The four years' complete disarmament plan of Khrushchev may be pure propaganda; or, what is more likely, it may be something the Soviet leaders would really like to see happen, but about the chances of which they have no illusions. In any case, it is an ideal which appeals to people all over the world.

The United States should show the world an ideal not less exciting but more realistic than "complete disarmament in four years." We should offer the world a broad and imaginative plan for world co-operation in all practically feasible areas. Science is one obvious field in which co-operation is possible, in fact, traditional. The International Geophysical Year showed how successful it can be, even between nations poles apart in their ideology and their political aims. Other projects of the same type are slowly getting under way, for example, international co-operation in the fight against certain diseases. It is deplorable that the American government is not in the lead in devising new plans of this type, that Soviet proposals for the pooling of efforts in the development of thermonuclear energy and in space exploration seem to find only grudging response in Washington. Khrushchev's accusation that America is dragging her feet in the international exchange of scientists may be unjustified; at the very least, it can be countered by evidence of delays and difficulties originating on the Soviet side. But to make such Soviet accusations at all possible, to insist on strict reciprocity, "one Soviet physiologist coming to the United States for three months in exchange for one American pathologist going to the Soviet Union for three months," is a sad demonstration of the failure of our government to boldly grasp the leadership. We should offer any number of Soviet students and scientists the opportunity to come to the United States, irrespective of the number of American scientists going to the Soviet Union. We—and the world—are bound to profit from pooling efforts in science, whether this pooling occurs on our own or on Russian soil!

In our time, the world has learned what progress in pure science and exploration of the unknown can mean in terms of acquisition by mankind of new power over the forces of nature. Atomic energy, undreamed of twenty years ago; space exploration, not so long ago a subject reserved to space fiction; the conquest of many diseases which only a short time ago used to ravage mankind unchecked: these are examples of the rapid conversion of advances in pure physics and biology to practical achievements able to benefit all men. The tradition of international co-operation in pure science must now be extended to applied science and technology. In this field, too, America and the West should show an imaginative leadership.[1] It is through this kind of practical co-operation that the peoples of different nations, and their political leaders, can gradually come to see in each other not enemies, not competitors, but partners in a common effort; and only through this feeling of partnership will they get rid of mutual apprehension. This seems a long and tortuous way to avoid dangers which now threaten our very existence; and it is difficult to give up the belief that these problems can be solved much more rapidly and directly simply by destroying the weapons or beating swords into plowshares. But this slow and hard approach leads to the true source of the danger—the antagonistic and self-centered attitudes of the nations and societies—instead of being directed merely at its manifestations—the acquisition by nations and societies of all kinds of murderous implements.

In one of my earlier articles in the *Bulletin of the Atomic Scientists,* I compared attempts to cure the world's sickness by disarmament to attempts to cure a patient by reducing his fever, without attacking the causes of the latter. In the case of a man as well as in that of mankind, attacking symptoms has only a limited effect and is often unsuccessful if the cause of these symptoms is a virulent infection or a persistent systemic disorder. The systemic disorder that causes the fever of the arms race is long-established and violent; to attack its symptoms and not its cause will not make mankind healthy.

[1] See the "Vienna Declaration" (pp. 251–57, in this book) for an expression of this idea.

The Dawn of a New Decade

January 1, 1960, marked the end of a decade. Looking back on the ten years which are now history, one is seized by a breath-taking thought: perhaps we have lived through a great turning point in the affairs of mankind. In 1921 the Russian poet Alexander Blok said, as he lay dying, that he could hear "the stormwinds of history" blowing over his head. The stormwinds of history have been blowing ever since 1914; by now, they have risen to a hurricane.

Several major developments have reached their climax in our time: the revolution of underprivileged classes against societies which failed to soften their social injustices; the revolution of underprivileged nations against empires which did not transform themselves fast enough into commonwealths of free nations; and—lending terrible urgency to these two upheavals—the great scientific and technological revolution, which has endowed mankind with an unheard-of capacity to destroy itself, or to reach out for new heights of material welfare everywhere.

On the day when the existence of the atom bomb was made known to the world, Robert Hutchins, then chancellor of the University of Chicago, declared, "The atom bomb calls for world government," and appointed a committee to write a world constitution. The constitution, elaborated at Hutchins' behest by Professor Borgese and his associates, is now all but forgotten; other elaborate plans for world reorganization (such as that described by Grenville Clark and Louis Sohn, to name but one among many) have left equally little imprint on world events. Even the much less ambitious "foot in the door" plan for international ownership and management of atomic energy, accepted in 1948 by all United Nations members except the Communist ones, now seems like a dream which only unworldly scientists could ever have taken seriously.

It seems that mankind is as far as ever from stable peace. The danger of war lurks around many corners. The dominant political passion of many peoples still is hatred for other peoples, desire for the fall of their political or economic systems, for the reduction of

From *Bulletin of the Atomic Scientists,* January, 1960. This article is reprinted with the permisson of Basic Books, Inc., which has included it in a recent anthology, *Seven Minutes to Midnight.*

their territories, or for their outright destruction—desires which cannot be realized except by war. Does this mean that "practical" men have been proved right—men who smugly asserted in 1945 (as some still assert now, although less smugly) that, bomb or no bomb, human history will remain the history of contests for power and that wars have always been, and will always remain, the inevitable climaxes of these contests?

Much contemporary evidence supports this skeptical view. The world scene is still ominously dark. Several countries of Eastern Europe remain under alien military rule, exercised through native ideological minorities but no less abhorrent for that to the majorities of these peoples. The minority that lost the civil war in China is still supported by outside powers in its hope of regaining the control of the country, where another minority that won the civil war has successfully established its domination. Armistice, and not peace, still rules in divided Korea and Vietnam; and not even an official cease-fire exists in the sea around Taiwan. And yet, whenever one of these smoldering power conflicts has flamed up in the last decade, the eruption has been quenched and the original situation, however patently absurd, has been re-established. The world map has been frozen by the universal fear of a great war. The Suez expedition was called off after the fighting was well under way—in fact, when it was almost over—although vital interests of two great powers had to be sacrificed. The troops that landed in Lebanon re-embarked; the threatened assault on Quemoy never went beyond an artillery barrage; the showdown around Berlin is being delayed indefinitely, despite the obvious local military superiority of one side.

This repeated frustration illuminated the decadence, in our time, of a diplomatic technique which had been successfully used in the past when threats of war carried conviction. Now, war threats and counterthreats have become bluffs and counterbluffs. Even if this change is for the better, it is not without new dangers. A power, threatened with a loss of face, may resort to force if its bluff is called, even if it did not seriously intend to do so in the first place. However, in the past decade, major powers have shown considerable caution in keeping the paths of retreat open for themselves and for their adversaries. We can hope that they will be even more cautious in the age of intercontinental missiles and missile-launching submarines, since these weapons make the possibility of a last-minute halt precarious and thus call for abstention from the practice of brinkmanship.

These are signs of the changing world, evidences of the impotence, in our time, of the diplomacy of the mailed fist, of gunboat demonstrations and marine landings, of the brandishing of ICBM's or atomic bomb-carrying airplanes. They are gratifying but not too reassuring. After all, one can say, abortive conflicts also preceded the first and the second world wars.

Little more reassurance can be derived from the decade of coat-tail diplomacy—from the dreary bargaining in London, Warsaw, and Geneva or from the easy triumphs (and occasional fiascoes) of the peregrinations of VIP's—be it Premier Khrushchev and Mr. Bulganin (remember these two garlanded with roses in India?), Vice-President Nixon, Premier Macmillan, or, most recently and most triumphantly, President Eisenhower. Nor are the evanescent exhilarations of the "spirits" of Geneva or Camp David to be taken seriously as signs of a new age. Equally dramatic was the encounter of Napoleon and Alexander I on a float on the Niemen, which was followed a few years later by Napoleon's invasion of Russia, not to stir up the painful memories of the "summit meetings" in Munich, Berchtesgaden, Yalta, and Potsdam. Relaxation of international tensions, softening of the long-rigid policies of Moscow and Washington, all this is for the good and should be eagerly fostered; but all these things have happened before in past power conflicts and never meant peace for more than a few years.

Where, then, can we see real signs of a new turn in human affairs? In the treaty for the demilitarization of the Antarctic and its permanent reservation for co-operative international research; in the resounding success of the International Geophysical Year, which pressed into service not only the full scientific resources of all nations but also their warships and rockets; in plans for other co-operative scientific efforts, particularly first steps toward world-wide pooling of space exploration; in the International Atomic Energy Agency; in the vigor lately displayed by specialized agencies of the United Nations devoted to world-wide relief and reconstruction—WHO and WFA, UNESCO and UNICEF, and the new UN Special Fund; in the work of various agencies of the United States and of the British Commonwealth in the rehabilitation of underdeveloped areas in Asia and Africa; in conversations between world scientists on the problems of war and peace, such as the several Pugwash meetings; in the obvious urge of peoples, in all parts of the world, to work together irrespective of ideological antagonisms and power conflicts between their governments. A new world of international co-operation is beginning to take shape under the frozen crust of the old world of self-centered nations deadlocked in power conflicts.

If this new spirit is fanned by those aware of the stake mankind has in its preservation, if the governments of the world will permit it to grow, then the trend of history may truly take a new direction in our time. Future generations may then come to see, in the years which now appear as an era of darkness, confusion, desperation, and deadly danger, the time when a break was first made with the age-long divisive tradition of mankind, when world community began to become a reality.

Three broad changes in man's awareness of human relations are pushing us in this hopeful direction: a change in man's relation to war, a change in man's attitude toward the rule of force, and a growing feeling of personal and national responsibility for the security and prosperity of mankind as a whole and not only of one's own country.

A generation ago, the belief in the naturalness of war, in the glory of victorious battles, in the rationality of the use of military power as tool of national policy, was universal. In a Russian "Student Calendar" which I used to buy every year before the First World War, there was a table showing the numbers of battles each country had fought in its history, with the comment, "France, the most civilized of all countries, has also engaged in the greatest number of battles." This attitude toward war is now dead. What was once the faith of a few exalted religious leaders and the reasoned conclusion of a few humanist philosophers—that war is evil and that the establishment of permanent peace must be the considered aim of mankind—has now become a common, everyday belief of men and women all over the world.

The same change has occurred in man's attitude toward the rule of force. The use of force in the maintenance of national strength was taken for granted as legitimate since the dawn of history. All empires have been founded on conquest; even now, while the historical empires of the European nations in Asia and Africa are crumbling, an attempt is being made in eastern Europe to stabilize a new empire, intended to be held together by a bond of common ideology but founded, like all empires before it, on military conquest. There is, however, a significant difference. In the past, no ideological justification was needed; the right of the stronger nations to rule over the weaker ones—England over Ireland or India, Russia over Poland, Austria over Italy, Japan over Manchuria, to take only the most recent examples—was not questioned and empires built on such conquest confidently hoped to last forever. Nobody believes now in the legitimacy or viability of the rule of powerful nations over weaker ones. If the Soviet rulers do not succeed in converting the East Germans, the Poles, the Hungarians, into loyal Communists, their domination of these countries will not last long—and the same is true of France's hope of making Algeria a stable part of France. We cannot foretell, in every given case, how the rule of force will be broken if it does not gain popular support; but we are now certain that no institution can survive for long if it is not accepted by men as natural and legitimate.

War and the rule of force always went together in the minds of men. Acquiescence in the recurrence of one and the permanence of the other has now disappeared from human minds; and this change has occurred in the short span of thirty years, between 1918 and 1948.

Each war in history has had its special cause, its aggressors and its victims, its heroes and its villains. Yet, there has been one common underlying cause of all wars—the existence of groups of mankind within which individuals have abandoned some of their power for the benefit of the community, while no ethical or legal restraints were imposed on their relations to other, similar groups. The purpose of each community was to assure for its members the greatest possible share of the limited wealth available on earth. In this, its interests were naturally opposed to the interests of other communities. One nation could not be rich except by others being poor, powerful except by others being weak.

Mankind still largely exists in this traditional framework. It still consists of self-centered sections pursuing their sectional interests as the *summum bonum;* but behind the continuing reality of a world divided into contending factions, into mutually hating and distrusting national, religious, and ideological units, there has begun to grow another reality—that of a humanity conscious of mutual involvement and responsibility of everybody for everybody, one for all, all for one.

Much too slowly—and yet, how rapidly, if considered in the context of history—the realization is spreading in America that assistance to less fortunate nations is the moral obligation of an economically strong country. Americans are beginning to understand that no nation has either a moral right, or the objective possibility, of surviving indefinitely as an island of prosperity in a sea of want. What was once the opinion of small groups of peculiar people, such as the Quakers, or of idealistic individuals, such as Albert Schweitzer, whom very few took seriously, is becoming in our time a common belief.

Vice-President Nixon suggests that the United States should accept the jurisdiction of a world court in its future international treaties. Mr. Khrushchev, as well as President Eisenhower and General de Gaulle, cautiously mentions the possibility of all advanced nations pooling their resources to assist the underdeveloped parts of the world. These words are not deeds, but they are portents. In the most cynical interpretation, the words of politicians suggest what they believe the public wants to hear, and only a few years ago such utterances would have meant the death by ridicule of an unwary American politician. Remember how Henry Wallace was accused of wanting to provide milk for Hottentot children? Was it not on that occasion that Mrs. Luce coined the ignominious term "globaloney"; and is it not ten years later, a Republican president, elected to put an end to such nonsense, who proclaims in New Delhi a "world war against hunger"?

In recent months, the United States government has begun to exercise pressure on its European allies to make them accept their share of the responsibility to aid underdeveloped nations; and if many in Europe think that this is none of their business, they hesitate to say so because this is not a proper thing to say in our time.

Of course, the abhorrence of war and the broadening feeling of responsibility for the well-being of all men are not due simply to the moral growth of the human race. As always in history, virtue is the child of necessity. In the past, devotion to one's people and country, renunciation of unlimited pursuit of selfish interests for the benefit of a racial or national community, could only become a general code of conduct when it was realized that by subjecting himself to this code an individual would improve his own chances for survival and the safety of his progeny.

In our time, the survival and prosperity of any individual or group is becoming more and more obviously tied up with the well-being and security of mankind as a whole. The selfish interests of these groups now call for the recognition of new ethical principles encompassing the whole of mankind. It is not a disparagement of the value of moral ideas in history to say that these ideas, always latent and expressed from the earliest ages by exceptional individuals, become powerful influences in social life when their immanent virtue and justice find expression in their practical importance for the well-being of men.

In the last year or two, while the arms race continued unabated, the contest between the West and the East has shifted to a new arena: to competition for the allegiance of the uncommitted parts of the world, gained by increasing production and by using this increase for world political aims. This is significant progress. Even if the immediate motive of political leadership in engaging in this new competition is the old quest for power, the idealistic ingredient in this effort (without which no national effort can succeed) is new. Competition in the development of military power is, by the nature of things, directed against somebody. The only aim of military power is to be able to destroy the military power of others. Competition in bringing about improvements in the well-being of other nations has a common positive aim and can easily become a step to co-operation. Three steel mills are being built in India, one with American help, one by Germany, and one by the Soviet Union; in an even more drastic case, in one town in central Asia, an electric power station was built by one great power and a streetcar system by another one. The logic of the situation calls for these efforts to become co-operative; and the most hopeful sign of our time is that this logic is beginning to prevail against the traditionalist blindness of those in the East who refuse to see in Western assistance programs anything but a drive for the enslavement of new nations by capitalist exploitation and of those in the West who see in Soviet assistance nothing but a particularly reprehensible technique of Communist subversion.

When, sixteen months ago in Vienna, a declaration was adopted by scientists of all countries,[1] calling not only for an end to wars but also for the co-operation of all nations, irrespective of their political and economic structure, in technological assistance to the less well-developed nations, this may have seemed a quixotic idea, which only men with no understanding of reality could cherish. Since then, this concept has started popping up in public discussions, including the pronouncements of responsible national leaders. It is not boundless optimism to hope that a few years from now, the concept of world-wide co-operation in the technical advancement of the under-developed parts of the world will be widely accepted.

The development of science and technology is rapidly changing the realities of human existence; one does not need to be a Marxist to say that this change in existence must entail changes in consciousness. The progress of scientific technology has given to fractions of mankind the capacity to destroy each other utterly and has thus made the historical concepts of international struggle for power obsolete; but human consciousness needs time to adjust itself to this new state of affairs, in which no security exists for any one nation except in the security of all of them. The same progress in scientific technology is converting a world of limited wealth, in which each nation (and each class within a nation) could be prosperous only at the expense of other nations or other social classes, into a world in which prosperity is available for all, if science and technology are pressed into the service of creating wealth; but the past experience of strict limitation of wealth, and the struggle for this wealth be-

[1] For the text of this declaration, see pp. 251–57.

tween nations and classes, is but slowly forgotten. When Premier Khrushchev was in America, he admitted that "the slaves of capitalism live well," in other words, that the capitalist system (which he believes to be inferior in effectiveness to a planned Communist economy) can produce enough wealth to keep everybody fairly prosperous. This is the kind of enlightenment that may permit a softening of the power conflict between the Soviet Union and the West, now exacerbated by exaggerated belief on both sides in the decisive importance for prosperity of this or that system of production and distribution of wealth.

These are the signs that a turning away from the path of traditional power policy is becoming psychologically possible. We do not doubt that, as of now, the mainstream of political events is still dominated by traditional thinking and by the inertia of established institutions. The outlines of a new world community are but vaguely discernible behind the traditional structure of divided humanity.

Nevertheless, in recognition of these new hopeful elements in the world picture, we are moving the "clock of doom" on the cover of the *Bulletin of the Atomic Scientists* a few minutes back from midnight. In doing so, we are not succumbing to a facile optimism engendered by a change in the climate of our diplomatic relations with the Soviet Union or to the exhilaration engendered by the personal contacts of the leaders of the great powers and their visits to different countries of the world. We want to express in this move our belief that a new cohesive force has entered the interplay of forces shaping the fate of mankind and is making the future of man a little less foreboding.

When, in the past, the *Bulletin* clock was moved forward closer to midnight, it was on the occasion of events—the first Soviet atom bomb, the first hydrogen bomb—symbolic of mankind's drift toward the abyss of a nuclear war. The recent advent of intercontinental missiles is another stage of the same drift; the forthcoming test of a French nuclear bomb in the Sahara, symbolic as it is of the beginning of the world-wide spread of nuclear weapons, will be another. No similar landmark can be pointed out indicating progress on the road to world community, but there has been, in recent years, an accumulation of facts and words which suggest that this hopeful trend is gathering force. The feeling seems justified that a turn of the road may have been reached, that mankind may have begun moving, however hesitantly, away from the dead end of its history.

7

Lessons of Cuba

The world has emerged from a breathtaking experience of brinkmanship. As on some past occasions, brinkmanship succeeded. The Soviet Union chose losing face rather than plunging the world—including itself—into destruction. Once again, a totalitarian system showed one of its few advantages—capacity for retreat, unhampered by public opinion.

The first superficial lesson some people may draw from this experience is that the Soviet Union will retreat when confronted with a real show of strength. Those who had always claimed that only the pusillanimity of Western leadership had permitted communism to spread as far as it did will say "We told you so." They will call for the same uncompromising spirit to be displayed in other areas of conflict.

This conclusion is a half-truth, and like all half-truths, it is dangerous. Foreseeable change in nuclear capacity of the West and the U.S.S.R. may in the future make the readiness of the Soviet Union to stand up to the challenge of nuclear war as real as was that of America during the Cuban conflict. If this comes to pass, it will dangerously increase the possibility that a future power confrontation would result in a war which nobody really wanted.

This is the true lesson of Cuba. It showed that the so-called equilibrium of mutual deterrence is uncomfortably unstable even under the present conditions; it can be expected to become more so with time. Instead of simply resolving to stand as fast in future confrontations as it did in October, 1962, the United States should search for a new world policy which would make repeated convulsions of power politics between the two great nuclear powers less likely, if not eliminate them altogether.

Why Did the Soviet Union Retreat?

The reason the Soviet Union yielded in the Cuban crisis had to do with the reason it had embarked in the first place on the adventure of emplacing rockets where they were certain to be located by American intelligence and to produce a violent reaction.

In the Cuban crisis, much was made of the difference between offensive and defensive weapons. This distinction would have been

From *Bulletin of the Atomic Scientists,* February, 1963.

meaningful if the Soviet weapons in Cuba were intended to defend Cuba from an enemy of approximately equal power; however, they were not put there for this purpose but as part of the Soviet strategic deployment in conflict with the United States. In this context, "defensive" and "offensive" are not significant terms. In our time, defense has been replaced between the great powers by deterrence; capacity to bring destruction to the enemy's country has taken the place of capacity to repulse attack on one's own country. The Soviet Union did claim that because the Soviet military posture is defensive, all Soviet weapons, including medium-range rockets in Cuba, are defensive; in the same way, the United States considers all American weapons everywhere in the world as defensive.

The distinction which does have meaning in deterrent strategy is the one between first-strike and second-strike weapons. Weapons openly deployed and susceptible to annihilation by an enemy who attacks first can be used only as first-strike weapons. Second-strike weapons are weapons hidden or protected so that they cannot be easily destroyed and can remain available for a retaliatory strike. The rockets in Cuba were typical first-strike weapons. If the Soviet military leaders were eager to place them in Cuba, they must have felt their present first-strike capacity to be in urgent need of a boost.

Prime Minister Khrushchev did boast that the Soviet Union possesses rockets (he called them "world rockets") able to hit America, not only over the short route via the North Pole, but also on roundabout trajectories. But since 1959, when the missile gap scare was proved baseless by U-2 reconnaissance, the number of Soviet long-range weapons has been known to be insufficient to substantially blunt an American retaliatory strike. Destruction of the enemy's retaliatory capacity is, however, the only reasonable aim of a nuclear first strike, because otherwise such a strike could only mean national murder *cum* suicide. The Soviet Union is incapable now of carrying out—or, what may be more significant, of plausibly threatening—a nuclear first strike. The United States, on the other hand, behaves as if it were capable now of a nuclear first strike, to which American rocket installations on the periphery of the Soviet Union provide a contribution. The Soviet Union has tried to scare the countries on whose territories these bases are located by threatening them with becoming first-priority nuclear targets in case of war, but this pressure has not succeeded.

Whether the United States in fact has an effective first-strike capacity depends, among other things, on the success of American intelligence in locating Soviet rocket bases. The Soviet Union cannot be certain of how much American intelligence knows, but it must have been alarmed when Secretary of Defense Robert S. McNamara proclaimed early in 1962 that the United States is switching targets for nuclear attack from population centers to military installations. An attempt has been made to present this switch as evidence of humanitarianism, but to the knowing it had another implication. Since an attack on the enemy's nuclear bases has to be a first strike, switching to these targets implied knowledge of their location and the power to destroy them. The Soviet analysts must have connected McNamara's statement with President Kennedy's statement to Joseph Alsop that the American leadership will not hesitate to use nuclear weapons first. This did not mean, as Soviet propaganda has tried to convince the world, that the United States was considering unprovoked nuclear aggression against the Soviet Union. But it did suggest that the United States was ready to answer a non-nuclear aggression with a nuclear counterattack.

The Soviet leadership must have concluded from these announcements that American leadership either possesses, or at least believes it possesses, an adequate first-strike capacity. This, and the weakness of the Soviet counterclaim of its own first-strike capacity, must have hampered Soviet diplomacy in its confrontation with the United States in Berlin and elsewhere. As long as threats of war are the stock in trade of diplomacy, possession of superior weapons remains a powerful diplomatic asset. The Soviet attempt to put first-strike weapons on Cuba in a hurry can be understood as an attempt to make the first-strike threats more nearly symmetrical.

American spokesmen described the placing of nuclear rockets on Cuba as an attempt to upset the strategic status quo in the world; this is correct if this status quo is taken to represent not a "balance of power" but the present asymmetry of first-strike capacity. With the failure of the Cuban gambit, the Soviet leadership must resign itself to continuing for a while the diplomatic war under the conditions of a one-sided first-strike threat. This calls for postponing diplomatic offensives, and recent statements of Eastern political leaders and the slogans of the October celebrations in Moscow suggest such a postponement.

INCREASED DANGER OF CONFRONTATIONS

These are then the special circumstances which may have accounted for the Soviet foray in Cuba and for its abandonment in the face of firm American opposition. Can we presume that the same asymmetry of nuclear threats will prevail in future confrontations? Such an assumption would be dangerous. If the arms race continues, the number, power, and precision of long-range missiles is bound to grow on both sides. Medium-range rocket bases on foreign soil will gradually lose their importance. It has been claimed that this has already happened. The claim was premature: the reluctance of American leadership to consider denuclearization of central Europe is as much evidence of this as was the Soviet eagerness to acquire medium-range bases in the Caribbean. But what was premature in 1960 or 1962 may become true in 1964 or 1966.

As the reliance on long-range megaton missiles (plus submarine-carried Polaris-type rockets) grows, two factors will work for equalization. The growth in absolute numbers will make the difference between them less decisive, and the "hardening" of the rocket emplacements (placing them underground or on submarines) will make the results of a first strike uncertain. In future political conflicts, the success of a threat of nuclear first strike may become less and less a matter of actual superiority in first-strike weapons and more and more one of readiness for ultimate, reckless brinkmanship.

Physics has the notion of "metastable states" in which a small statistical fluctuation is sufficient to cause massive condensation or crystallization. The cold war is such a metastable state; the Cuban conflict was a fluctuation which nearly destroyed its stability. When the state of approximately equal first-strike capacity is reached between the West and the U.S.S.R., fluctuations will become even more dangerous, because each side will be tempted to go closer and closer to the brink in the hope that the other will "chicken out" first.

Alternatives of Military Policy

It may be that we have no choice but to continue the deadly game of power politics and the arms race with the Soviet Union until one side collapses under its strains or both go down in a nuclear catastrophe. (Cold war is simply a new expression for power politics, appropriate to an era when the buildup of military strength cannot be postponed until after the outbreak of a war.) But is there no alternative on either the military or the political plane?

One alternative military policy has been proposed. American scientists (for example, Richard Leghorn in the June, 1958, issue of the *Bulletin of the Atomic Scientists*) have suggested that both sides should renounce first-strike nuclear weapons, but be permitted to maintain a number of second-strike weapons—hardened, land-based missiles, or submarine-carried rockets—sufficient to assure that no attack on it could be made without intolerable retaliation. This plan has received the inadequate name of "arms control." It was presented to Soviet scientists at several Pugwash Conferences on Science and World Affairs (COSWA), but the reaction on the Soviet side has been to denounce it as a subterfuge to avoid nuclear disarmament.

Recently, the Soviet representative in the disarmament talks was quoted as suggesting that the first stage of the Soviet disarmament program, which called for complete elimination of vehicles for the delivery of nuclear weapons, could be amended to permit both sides to retain an agreed-upon small number of such vehicles. That may mean Soviet recognition that their present inferiority in first-strike weapons would make the arms control plan advantageous to them. But would American leadership be willing to consider this proposition with favor now (in contrast to the period of the missile gap scare), in face of the advantage which American superiority in first-strike power has brought in the Cuban conflict? This may be doubted; yet, in view of the probable long-range developments, mutual abandonment of first-strike capacity may still be a good idea, despite the fact that it would mean accelerating the advent of a true nuclear balance. This balanced state may be less dangerous to both sides than the one to which the present development—a sky's-the-limit race for predominance in first-strike capacity—is likely to lead.

Accepting symmetry in nuclear power would of course also call for approximate equalization of non-nuclear military strength. This is within the capacity of the Western alliance, and it seems that it is being actively pursued by American leadership but resisted by Britain, France, and Germany, because it would require much costlier military efforts on their part.

The third alternative of military policy—the one to which both sides are now committed in theory, while they continue the policy of unlimited arms race in practice—is disarmament. Far-reaching or complete disarmament would also mean equalization of military power—equalization on the zero level instead of on a certain intermediate level, as in the arms control plan.

The three alternatives of military policy are thus: all-out arms race, stabilized deterrence (arms control), or disarmament.

Alternatives of Foreign Policy

But military policy is inescapably tied to foreign policy and cannot be divorced from it. A military policy aimed at maintaining an asymmetry of nuclear striking power is the logical concomitant of a foreign policy based on assumption of an asymmetry of political status, which the West has adopted since the Soviet revolution.

The world has been divided, from the Western point of view, into legitimate and essentially defensive nations on one side and an ideologically and militarily aggressive power on the other side. The military strength of the first camp had to be superior, just as a police force, protecting citizens against unruly elements, must be superior to that of any potential violator of order. The concept of a *cordon sanitaire,* created by France and England in Eastern Europe after the First World War, has not quite been given up—despite the intermezzo of the Second World War. The place of the *cordon sanitaire,* overrun and converted into a Soviet bastion in the center of Europe, has been taken by the threat of American nuclear weapons.

The concept of unequal political status is incompatible with any military policy deliberately accepting an equality of military status, on whatever level it may be; whereas acceptance of political equality is now incompatible with the continuation of the power contest and of the arms race. An equality of political status could be accepted, for example, between the Triple Alliance and the Entente Cordiale before 1914, despite the power struggle between them, because before the advent of nuclear weapons, not only the all-out arms race, but also the conversion of a political rival into a deadly foe, of a legitimate competitor into the embodiment of all evil, could be held in abeyance until the outbreak of a war. In a nuclear age, this has become impossible.

The revolution in military technology has converted the leisurely arms race of the past into the frenzied arms race of our time; it has also made political "armistices" impossible. The alternative is now between the continuation of the cold war and the establishment of a permanent, stable peace. Stable peace means, however, the end of power politics—of politics based on the assumption of fundamentally contradictory political interests of the competing powers. The machinery of power politics has become crippled since the development of nuclear weapons, because without plausible war threats, power politics becomes a game of bluffs and counterbluffs, of propaganda and counterpropaganda, of mutual subversion and intrigue, continuously threatening to explode into an all-destructive war that nobody wants. But, with its machinery crippled, power politics still goes on because it is the only kind of politics nations and their leaders know how to conduct.

Could power politics be continued under the conditions of arms control? Leo Szilard made an attempt to devise a rational system of power politics suitable for these conditions (December, 1961, issue of the *Bulletin of the Atomic Scientists*). He arrived at a scheme of "city-barter," in which the achievement of a power-political aim of one side would remain possible, provided this side declared itself willing to accept the destruction of one or several of its cities in exchange for a city or cities of approximately the same size of the opposing side. This grotesque scheme, resulting from a thoughtful attempt to combine the existence of nuclear weapons (even if only second-strike weapons) with the continuation of power politics, suggests that reasonable power politics would be as impossible in an era of arms control as under the conditions of continued arms race, and would degenerate in practice into the same game of bluffs, propaganda, subversion, and brinkmanship.

Continuation of traditional power politics under the conditions of disarmament appears even more contradictory. Recently, even the Soviet Union has shown willingness to discuss the organization of peace in a disarmed world, because without a system of international adjudication, and its enforcement by an international police force, no extent of disarmament could prevent wars. War would start on a very low level of military organization, but once it had started the warring sides would have no choice but to build up their military establishment at maximum speed—even re-creating nuclear weapons might not take long. The know-how of scientific weaponeering, once acquired, will remain with mankind for all of its future.

A New Political Doctrine

To make possible a military policy other than that of an all-out arms race, a new political doctrine must be elaborated based on acceptance of an equality in political status.

Western reluctance to accept the Soviet Union as a legitimate partner in a world-wide political system is based on the bitter experiences of the Soviet application of military power in pursuit of its political aims—from the Polish invasion in 1919 to the communization of Eastern Europe in 1946, not to mention the invasion of South Korea and the attack on Finland. These experiences cannot be forgotten; but neither can it be denied that the policy of "containing" the Communist nations of Eastern Europe by superior military power—meaning, as of 1962, superior first-strike nuclear threat—and waiting for their evolutionary or revolutionary conversion into political systems more acceptable to the West is becoming increasingly dangerous. Since the containment policy was first conceived, the Soviet Union has acquired the capacity to destroy at will all Western Europe and is in the process of acquiring the capacity to similarly destroy America. This development calls for a search for a new political doctrine rather than simply for a stubborn decision to stand firm on the edge of the precipice.

Of course, this search may fail if the Soviet Union proves incapable of changing its by now "traditional" attitude of coexistence on paper but expansion in practice whenever possible. But is this change out of the question? In recent years, the policy of the Soviet Union appears to be dominated more and more by the desire for an equal status in world affairs rather than by a desire to spread its power beyond its present limits. The equality which the Soviet regime now wants is not owed to it by any right or rule of fair play, since it includes legitimization of an arbitrary rule, including rule over countries where more truly independent and more satisfactory ways of life are very recent recollections. The ambition of the Soviet government seems to be to stabilize the present political setup in its sphere of influence, to give it legitimate status equal to that of the Western alliance and proportional power in the councils of the world.

Even Soviet attempts to change the status of West Berlin appear intended to uphold the shaky East German state rather than to seek an opening for further expansion into Western Europe. The same may be said of Soviet ambitions in the United Nations. The Soviet "troika" proposal must be considered as part of a campaign for equality rather than as a drive for domination of the world organization. It is unacceptable as long as the basic political aims of the two sides are incompatible, because under these conditions its acceptance would paralyze the UN executive. It would become discutable, as coalition governments become in time of national emergency, if a common political doctrine, overriding contradictory interests, could be established.

It has often been asked whether the policy of the Soviet Union is the traditional policy of an expanding empire or an ideological crusade to spread the Communist gospel. The answer is that it is both. But history teaches us that every revolutionary power, be it religious or ideological, goes through a peak of crusading spirit and gradually settles down—without giving up its religious allegiance or its ideological credo—to de facto tolerance of other creeds. The Germanic invaders of the Roman Empire, the Moslem invaders of Christian Europe, the libertarian revolutionaries of the eighteenth and nineteenth centuries—all of them have become with time stable powers with vested interests in the preservation of the status quo, safeguarding their individual or collective property and seeking external security for their plans for internal growth.

The question is whether the Soviet Union, with its growing national assets, is now in a stage where interest in its own preservation is more powerful than interest in expansion and whether Western policy could do something to assist in this transformation. There is no way of finding an answer to this question except by trying to engage the Soviet Union in the discussion of a stable peace, not of a temporary armistic in a continuing conflict of political interests. The attack of the People's Republic of China on India and the comparison between the attitudes of China and the Soviet Union in the Cuban crisis have thrown into sudden relief the difference between a revolution in the reckless crusading stage and a revolution in the process of settling down.

The policy of stable accommodation between equal partners presumes that these partners have common interests which weigh more heavily than their contradictory or mutually exclusive interests. Can this be true of the relations between the Soviet Union and the West? Both sides do have a common interest in avoiding mutual destruction in a nuclear war; but this negative common interest would hardly suffice if their positive aims were hopelessly contradictory.

Two areas of common positive interest exist, however: the interest in stabilizing the political map of the world before China becomes strong enough to jeopardize this stabilization, and the interest in co-operative utilization of science and technology for the benefit of all nations.

POLITICAL STABILIZATION

Both the Soviet Union and the West are powers whose interest in world stability is—or rationally should be—stronger than their interest in gaining additional areas of military or ideological domination. The transformation of the Soviet Union into a power interested in stabilization rather than expansion is being speeded up by the growth of the new revolutionary force of China. Even at the present time, it is not certain whether the combined efforts of the United States and the U.S.S.R. could prevent China from acquiring a nuclear capacity. With China a full-fledged atomic power and the West and the Soviet Union still at loggerheads, God help both of them!

It has often been said that the only way to end a long-standing enmity between two nations is to find a common enemy. Without suggesting an American-Soviet alliance for the containment of China (which would be both wrong and impossible), one can see that China provides a strong argument for the West and the Soviet

Union to bury the hatchet between them and to establish a stable system of international relations as long as they can do it with relative freedom from Chinese interference. To avoid the worst, both the West and the Soviet Union must be willing to freeze existing boundaries and the existing political status in all parts of the world over which they have a measure of control. Both should be willing to forego attempts to change this status, however tempting —and however justifiable from the point of view of history or international law—these changes may appear.

In their actual political behavior in the last ten years, both sides have acted in accordance with this rule. The West did not succumb to the temptation to break into the Soviet political system at the time of the Hungarian revolution; both sides have agreed on a buffer status for Laos; the Soviet Union has given up a military base in the western hemisphere; the Anglo-French excursion to Suez, which threatened to escalate into a major war, was called off under the combined pressure of the United States and the U.S.S.R. Is it not time now for these separate lessons of experience to congeal into a piece of general political wisdom—a new political doctrine acceptable to both sides? Could not the United States approach the Soviet Union—as one power interested in world stability to another—with a proposal to co-operate in a world-wide stabilization policy?

Stabilizing the present political setup in the world could not imply, on the Western side, giving up sympathy for the aspirations of nations now deprived of free political decision—first of all, German aspiration for national unity. But to actively foster these aspirations with no possibility of using military power in their support would mean keeping the world on the brink of nuclear war for an indefinite time. We should make up our minds that we expect these wrongs to be righted by internal developments in this part of the world and not by military victory or external subversion. The same applies to the Soviet sympathy for potentially communist, nationalist, and socialist movements in former colonial countries. This sympathy will not disappear, but active support of these struggles will have to end, because it would make stabilization of relations between the West and the Soviet Union impossible and the danger of war permanent—as the Cuban experience has shown. The Soviet Union, too, must choose its paramount goal—preservation and internal growth or expansion under continuous threat of annihilation.

The first stabilization step on our side will have to be recognition of the present frontier between Poland and Germany. This frontier has ceased to be a problem of self-determination of peoples. However much our sense of justice may revolt against the solution of territorial controversies by pushing whole populations out of lands where they have lived for generations, we must not forget that Germany herself has initiated violent changes on Europe's ethnographic map, so that suffering the consequences of this initiative on its own national body is, in a sense, a deserved retribution.

A similar situation prevails in Israel, and the same principle of stabilization should be applied there. The Soviet Union will have to give up its non-recognition of the territorial status quo which, at present, is a useful tool in its struggle for influence in the Arab world. Territorial stabilization will not satisfy either the Germans or the Arabs, but if it is agreed upon and guaranteed by the West and by the Soviet Union, it will have to be accepted—at least as long as China can assist dissatisfied nationalists with nothing but resolutions of solidarity.

INTERNATIONAL CO-OPERATION

It is worth noting that in all recent negotiations between the East and the West one point, and often one point only, has ended in an agreement: improving cultural and scientific exchanges. Of all national activities, those in the cultural and scientific area are the least controversial and the most suitable for international co-operation. Exchanges in the fields of art, theater, and music and increased personal travel can counteract psychological alienation fostered by political separation of the two great camps. Exchanges and co-operation in pure and applied science, in addition to serving the same purpose, can have a much greater significance. In our time, science has ceased to be a socially unimportant occupation of small groups of intellectuals and has become an important national activity, consuming a sizable fraction of the national budgets; its direct relation to the health, prosperity, and technical status of a nation is being rapidly recognized, even by most conservative political leaders. Science, however, is intrinsically the first common enterprise of mankind, and increasing national interest and financial support for this activity means increasing the investment of all nations in a common enterprise—even if much of this investment is being justified by the competition in the arms race and the search for national prestige in the race to the moon.

A few years ago, it would have been considered preposterous to speak of international co-operation in scientific research as an important item in international relations. It was looked up as a little idyllic byplay in the great tragic spectacle of political and military contests between the great—and intrinsically antagonistic—powers of the world. However, since science has made the traditional mechanism of military resolution of power contests between nations rationally unacceptable, constructive international co-operation in science has grown in importance. If mankind is condemned by the existence of nuclear weapons to live in permanent peace, giving up what was long considered the highest and most exhilarating national experience—fighting for national or ideological or religious glory—it will have to learn how to enjoy and utilize this enforced peace for the greater benefit of all men. Scientific collaboration offers the most promising field for the retraining of mankind for life in a single community. There is little doubt that the Soviet Union is ripe for participation in such international efforts. In recent years, many initiatives in international co-operation in space research, oceanography, and other areas, have come from Soviet scientists. While Soviet scientists cannot always guarantee wholehearted co-operation of their government, they do constitute an important enough element of the new Soviet elite to hope that an imaginative Western initiative in a large-scale, comprehensive program of international co-operation in science will have a good chance of acceptance by the Soviet Union.

The national budgets in the field of science and technology are quite large—even if one forgets for the time being that military expenditures, as well as expenditures on technical assistance to underdeveloped countries, are also largely expenditures on research and development. Spending a considerable part of them on internationally sponsored programs would not be ineffective in creating a consciousness of the large areas of common interest and common effort that exist among the nations of the West and the U.S.S.R. It has often been suggested that a common flight of a Soviet and American cosmonaut would do more to bring together the two now alienated fractions of western civilization than could any political agreement or even an agreement on disarmament.

It is clear that at the present time the situation between the western world and China is quite different. In science, as in military technology, the Soviet Union and the Atlantic community are "have nations," whereas China and its satellites are "have-nots" and, as such, psychologically unready for co-operation.

However, in technical collaboration as in the stabilization of the political status quo the new departures in international relations made necessary by science are, like progress in science itself, neither temporary nor restricted in space. The common interest of the West and of the Soviet Union in stabilization and co-operation is supported, at the present moment, by a common apprehension of a new revolutionary force in the East. However, in contrast to past times when enmities between two nations could be temporarily replaced by co-operation in the face of a new competitor (the conversion of Germany and Japan into allies after the Second World War is a recent experience), the factors which now call for common Western and Soviet efforts are not temporary but permanent; if mankind is to survive, the new revolutionary power in the East will have to join in the same partnership sooner or later.

IN SUMMARY

The successful liquidation of the Cuban foray was due to the present imbalance in first-strike nuclear power between the United States and the U.S.S.R. If the all-out arms race continues, the asymmetric conditions which made the West successful in this confrontation are likely to be upset, and brinkmanship is likely to become more stubborn and more dangerous.

This calls for exploration of new political and military doctrines which could forestall these increasingly dangerous confrontations between the two nuclear superpowers. In the search for such a doctrine, it should be taken into consideration that military programs and programs of foreign policy cannot be divorced from each other. The present Western foreign policy, based on historically justified reluctance to consider the Soviet Union an equal partner in the political field, logically permits no other military policy but that of "winning" the arms race. Military policies of arms control or disarmament, both of which aim at equality of nuclear capacity on both sides, cannot be realistically pursued except if coupled with a new political doctrine. This doctrine must be based on acceptance of equal interest of the Soviet Union not only in the avoidance of a nuclear war but also in the political stabilization of the world.

The Soviet Union is evolving from a revolutionary, aggressive "have-not" power into a power interested in preservation more than in expansion. In what stage of this progress it is now, we do not know; but the existence of a new revolutionary "have-not" power in China accelerates this evolution. Therefore an American political initiative aimed at over-all stabilization by agreement with the Soviet Union is worth a try. A comprehensive and imaginative program of international co-operation in pure and applied science —including health, agriculture, space technology, and assistance to underdeveloped countries—could be an important adjunct to the policy of political stabilization, creating on both sides the consciousness of a wide area of common interests.

Finally, because of the universal character of scientific progress, an association of the West and the Soviet Union in a common political and scientific program could not be a temporary alliance directed against China as a common competitor but would have to be aimed at the gradual involvement of China itself.

8

Things To Come—Now (1962)

A Proviso

In 1939 and 1940, at the beginning of the Second World War, it proved possible to give rather precise predictions of the course and outcome of the war.[1] Contrary to what most people think, the course of this war was largely predetermined at the moment it started because the forces and the masses which were going to decide the outcome were given—at least, approximately—and no important new ones were likely to arise—or, in fact, did arise—before the end of the war. It is much more difficult to make predictions in peacetime, when masses are in slow fermentation and forces continuously change their intensity and direction.

In one respect, the variety of possible future developments is now narrower than it was in periods between wars before 1945. One "degree of freedom" has been lost in world politics—the freedom to start a war in pursuit of rational political aims. This restriction has been amply demonstrated by events since 1945—by the aborted Suez invasion, by the inaction of the West in the face of the Hungarian uprising, by the Korean War ending where it started, by the Soviet rockets being withdrawn from Cuba, by the West's toleration of the partition of Germany and by Moscow's toleration of the thorn in its flesh which is West Berlin, and finally, by the Chinese withdrawal from most of the territory conquered in the Himalayan campaign.

The reason was in all cases the same: the prospect of an intolerably destructive nuclear war or, at least, of an "escalation" of the local conflict, the end of which could not be foreseen. The latter may have been the case in the latest war that did not come off—the Chinese attack on India. The full reasons that had moved the Chinese first to attack and then to withdraw remain dark; but if the Chinese troops had carried their campaign deeper into India, tenuously supported through the Himalayan passes, they undoubtedly would have risked becoming mired in a wide, hostile country, facing increasing Western intervention, growing distress at home, and danger of invasion from Taiwan, with no certainty that the Soviet Union would render them substantial assistance, thus risking a general nuclear war. Thus, even a war between two non-

[1] See "Things To Come—Then (1939–41)," pp. 5–11 of this book.

nuclear powers in the farthest corner of Asia was called off lest it get out of hand.

Between 1939 and 1945 events had to run their way within the rigid framework of war; alternatives to fighting had to remain in abeyance. The present situation is, in a sense, complementary: of all theoretically available ways, only those are rationally open which exclude war. This still leaves open threats of violence, "brinkmanship," local application of force in guerilla campaigns, support of internal subversion—everything "short of war." Because of this restriction of rational behavior—and the compulsion to remain rational is more powerful now than it was ever before—one can expect indefinite prolongation of highly unstable situations, such as that in West Berlin, which would have rapidly led to a war in the past. Such situations will resolve themselves in the long run, but by the general trend of world events and not by drastic decisions of exasperated military or political leaders.

The same factors that permit lasting metastable situations give a new significance to accidents, truly unpredictable small events with disproportionally large consequences. We know in physics that every macroscopic assembly of particles undergoes continuous fluctuations. These are unimportant when the system as a whole is in equilibrium or in slow transformation. But if the assembly is metastable—a physical chemist would say, if only a small "activation barrier" protects it from violent change—a fluctuation may suffice to overcome this barrier. In the 1920's, an underground storage tank, supposedly containing a fertilizer, ammonium nitrate, blew up in Germany with disastrous consequences to the nearby city of Ludwigshafen. It was whispered at that time that the storage room contained a high explosive concealed from Allied supervision. I do not know whether this suspicion was later confirmed; but the alternative hypothesis was that the allegedly "stable" fertilizer blew up in consequence of a local fluctuation of temperature. Theoretically, this *could* happen—once in a hundred, a thousand, or a million years; it is an extremely *unlikely* but not an *impossible* event.

The same is true of human affairs, in which the behavior of nations sums up the behavior of millions of individuals. The peculiarity of our time is that, more than ever before, a fluctuation, perhaps caused by the behavior of only a few individuals, can radically affect world events. In the past, unpredictable actions of individuals often appeared to have had a decisive effect on world history; but the farther the events receded into the past, the clearer it became that history was shaped by deeper and more ineluctable forces than the will or whim of an individual.

Be that as it may, the accidents which can now upset all predictions are of an altogether different kind from the past deeds of national or ideological heroes. They do not consist in the possible unleashing, by new leaders, of new political, ideological, or religious movements that change the behavior of peoples. They are literally "accidents." One can predict, with some degree of certainty, what the American or the Soviet *leadership* will do under foreseeable circumstances; but one cannot predict for certain whether the commander of a missile crew will suffer a nervous breakdown or a staff analyst misread an electronic record. Yet, such a mistake could now snowball into a catastrophe out of all proportion to its cause, as an accidental fluctuation of temperature can cause the explosion of a metastable chemical system.

The error of a subaltern will be in most cases corrected by higher authorities; but what about a rash action of these authorities themselves? We have seen in our times how individuals such as Hitler or Stalin, with obsessive ideas and a fanaticism close to madness, can gain decisive political influence and command unquestioned obedience of whole nations. Even in democratic countries, where a man of such a mental make-up could not reach the position of supreme authority, momentary mental confusion of an otherwise rational leader remains possible; staff consultations, which would normally correct errors in the judgment of a single man, may be practically impossible under the conditions of a five-minute warning. The strain posed on the President in the face of such a warning could be well-nigh intolerable and lead to a rash decision even on the part of an otherwise deliberate man.

Another possibility, which seems fantastic but cannot be entirely excluded has been suggested: the catastrophe of a nuclear war could be provoked by a small nation, or even a group of fanatics, who had come into possession of a nuclear weapon, and exploded it in such a way as to suggest attack by one of the major powers. The authorship of such an explosion will not be easy to establish if the bomb is planted by an agent or launched from a submarine. There is a story about an old lady with a cute little fox terrier, who left her pet in the hall of a hotel tied up among ferocious-looking Newfoundlands, Alsatians, and Great Danes. As soon as the lady left, the terrier bit his neighbor—a Dane—then jumped over him and bit an Alsatian on the other side. Soon the big animals were at each other's throats. When the lady rushed back, she found her pet whining innocently in a corner, while the other dogs were fighting furiously.

True, such a catalytic war is even less probable than an accidental one. But according to the theory of probability, the longer one waits, the more likely it is that an improbable thing will happen. How long is it safe to wait on the assumption that a storage tank of ammonium nitrate will not blow up? How long is it safe to assume that the metastable situation of the world abundantly stocked with nuclear weapons will not be upset by an accident, by the design of a small group, or the provocation of a single man?

Predicting the fate of mankind in the next decades is like predicting the fate of a drunkard who is picking his way along the edge of a precipice, swaying right and left as he goes. If this swaying does not bring him over the edge, one can figure out where and when he will arrive. What will happen if mankind does stumble over the edge of the nuclear precipice, defies all reasonable prediction, even of such intrepid crystal gazers as Herman Kahn or Nevil Shute?

All this is said here to qualify in advance the attempt to predict future events by evaluating the main masses existing and the main forces operating in the world in 1962. All such predictions must be preceded by the proviso, "This is what is likely to happen if no nuclear war breaks out by accident or by willful act of a small group of people."[2]

Another difficulty of predicting the future now is the uncertainty of the *time scale*. Under the compulsion of war, resistance to change is broken down and events develop at their full intrinsically possible rate. The brakes imposed by the American political system were largely released in wartime. The one-man majority by which the introduction of military service in the United States was decided before the Pearl Harbor attack was replaced by a one-woman minority in the vote on a declaration of war after it! In peacetime, even changes widely supported by the people, and consistent with the general trend of historic development, may be delayed for years by general inertia, or by willful opposition of a small minority —witness the civil rights legislation in the United States. Under the pressure of war, Churchill was able to offer common citizenship to England and France in 1940; in peace, European defense integration was voted down by the French parliament, and integration

[2] It is worth considering that even a closely averted nuclear catastrophe may have an unpredictable effect. An "anonymous" nuclear explosion, prevented by some safety device (such as the recently proposed direct communication line between the White House and the Kremlin) from unleashing a war, could radically change the mood of nations; it could open the door to the acceptance of a world-policing authority, unthinkable before the incident. (Perhaps, a nuclear explosion "of unknown origin" would be an effective way to break the inertia now frustrating disarmament negotiations!)

of Great Britain into the European Common Market remains at this writing highly uncertain. Because of this difference, the predictions of things to come in the Second World War could turn out to be approximately correct not only in their *content* but also in their *schedule;* whereas in 1962, a prediction of things to come has to be uncertain not only as to *what* may happen but also as to *when* it might happen—and this often also means whether it will have the chance of happening at all!

A final factor is the possibility of radical innovations in technology. The one wartime development that could have upset the 1939 predictions was the perfection of the atom bomb. Earlier wars had been fought with weapons known long before their outbreak. In the Second World War, two new weapons appeared during the war itself: radar, which blunted the German air attack on England, and the atom bomb, which catalyzed the capitulation of Japan. Of these two, radar was already "in the cards" when war started in 1939; but the atom bomb had been, in 1939, only a glimmer in the eyes of a few physicists. If the atom bomb had become ready much before 1945, it would have changed the course of the war; and if it had appeared on the German, and not on the Allied, side, it could have changed its outcome.

Since 1945, one new technological development has changed the world-wide military picture—the long-range rocket. The H-bomb merely gave greater emphasis to the already deadly threat of the A-bomb; whether an explosive power of ten megatons or of "only" one hundred kilotons TNT can be delivered in a single package is of no decisive importance; but that this package can be delivered half-way around the globe within a few minutes does create a new situation indeed.

In the following discussion, the technologically induced military deadlock is taken to be a lasting state of affairs, not to be upset, within the time to which the discussion applies—say, the next twenty-five years—by innovations. In fact, the discovery of new offensive weapons could do no more than emphasize the deadlock. The discovery of defensive weapons, capable of defeating a nuclear missile attack, could upset it, but the assumption here is that such discoveries are unlikely, despite the much-advertised Soviet and American successes in intercepting missiles by missiles. There is all the difference in the world between intercepting a single missile fired at a known time from a known place in a known direction and intercepting all, or almost all, missiles in an unexpected massive salvo, particularly if it is accompanied by the launching of decoys and interference with radio communications.

THE MAIN MASSES

Let us now consider the main masses and the principal trends at work in the world today. The big masses of humanity are four. First is Western Europe and North America—a total of a half-a-billion men with the most highly developed science and industry and with access to the largest stores of natural resources of the world. These are peoples who have generated modern science and who still carry most of its development, both in acquisition of new knowledge and in its application to the needs of mankind. The second mass is the Soviet Union and its European allies—a total of about one-third of a billion people historically closely related to and only recently (and incompletely) severed from the Western world, to which they belong by their history and by their past participation in technical and scientific progress. General de Gaulle likes to say that Europe extends from Brest to the Urals; but Europe, in the sense of common civilization and common history, extends around the globe, from Alaska to Kamchatka; the so-called Atlantic community is a large part but not the whole of it.

The third mass of humanity is formed by six hundred million Chinese, with their own civilization going back thousands of years. When Churchill called the Soviet Union a "riddle wrapped inside an enigma," he coined a description which applies to China much more than to the Soviet Union. The Soviet Union is, in essence, Russia, a part and parcel of Western civilization, only clothed in new (Western-made) clothes of a Communist society. In China, it is the race itself, and not only its present Communist incarnation, which is alien to the Western mind. When we were schoolboys, "world history" was the history of European peoples, interrupted by the Mongol, Turkish, and Arab invasions, which could as well have come from Mars, so little we knew (or cared to know) about the history and civilizations which lay behind these eruptions. Much has been written since about these invaders, but familiarity with the East remains, at best, cerebral, with no feeling for the past or future of the close to a billion people of southeast Asia. Kaiser Wilhelm once pretended to speak for all Europe when he denounced the "yellow peril" hanging over its future. We know now that the yellow peoples are not a swarm of human locusts ready to consume the thriving fields and cities of the West but carriers of equally old, if not older, cultures which have known periods of flourishing, as well as periods of decay, and are capable of creative efforts in material civilization, art, science, and political organization not less impressive than those of the white race; but our understanding of the motivations and yearnings of these peoples remains dim. Communists used to think that they had a key to the understanding of the behavior of all peoples in the consideration of the stage of their economic development on a universally valid scale. I am not so sure that the Soviet leaders are now satisfied with their understanding of China and its ways and not worried about the turns which China's history may take in its Communist phase.

The fourth great mass of humanity—also of the order of one billion—is made up of what is now described as "uncommitted nations." These range from the primitive inhabitants of the jungles of Africa or Asia to the sophisticated city people of India and Latin America. Some of them now feel themselves part of the Western world, others are its reluctant allies, still others invoke a "plague on both their houses," and some tend to support the Communist bloc. Their character and national endowments differ widely; but history has brought them together in a community of "have-nots," latecomers to the feast in which the wealth of the world had been divided between nations first to reach technological maturity. Their common trend is to struggle for independence and viability—both political and economic—and to try not to become pawns in the conflict between the communistic East and the libertarian West. Some nations in this group may be temporarily considered as safely in the Western camp (but we have just seen how precarious this "safety" is in the example of Pakistan!), others, such as Cuba, as having tied themselves irrevocably to the Communist bloc. But only two kinds of truly stable ties exist—the voluntary ties of common interests and the compulsory ties of military occupation. Neither of the two exist between Karachi and Washington or between Havana and Moscow.

THE MAIN TRENDS IN THE WEST

What are the main trends at work in the Western world? They appear rather encouraging. The dark future extrapolated by Marx and Engels for the capitalist economies has been averted, perhaps permanently, and certainly for some time to come, by the breakthroughs of the New Deal in the United States, by the creation of the welfare state in Scandinavia and in Great Britain, and by more recent, similar developments in France, Italy, and Germany. These reforms have converted industrial workers, the numerically largest class in technologically advanced nations—the class which, according to Marx and Engels, could have no ties of loyalty to the existing society and state—into an increasingly prosperous and increasingly conservative part of the nation.

Stevan Dedijer said that capitalism was saved, in the nick of time, by the scientific revolution. This development can be traced to Henry Ford's assembly line. It made mass production more advantageous to the producer than the making of a small number of luxury items. Large-scale production required a large market and so created a vested interest of the entrepreneurs in the buying capacity of the laborers. The continuation of a society in which the majority toiled to produce small amounts of goods accessible only to a small minority became impossible. An industry producing eight million cars a year could not survive if only a few thousand people in the country could afford to buy a car; agriculture able to produce more grain, butter, meat, and eggs than the whole population could consume without overeating could not exist if the majority of people were reduced to a starvation diet. The same capacity for increased production now makes the foremost Western nations concerned with the increase of the buying power of the underdeveloped continents—Africa, Asia, and South America. What began as a drive to exploit the masses of this "colonial" world, by selling them cheap industrial goods in exchange for valuable raw materials, is being transformed, by ineluctable technological trends, into a drive to convert these masses into consumers of high-grade finished goods and machinery.

It is now possible to see a general trend which the scientific revolution, through the rise of productivity it engenders, imposes on all Western societies. It is the trend toward a type of society which not only provides assistance for the education, health, and old-age needs of its citizens but is also concerned with wide diffusion of consuming power needed to use fully the growing productive capacities of an industrialized and automated economy. Communist theoreticians grew up—and think they still live—in the time of frenzied competition for external markets between capitalist economies handicapped by insufficient domestic buying capacity. They predicted that this competition would inevitably lead to self-destroying wars among capitalist nations. After the end of the Second World War, Stalin confidently expected the renewal of such internecine capitalist wars. Communist theorists still seize on every sign of sharpened market competition, be it between West Germany and England, between the Common Market and the British Commonwealth, or above all, between Europe and the United States, hoping that it will divide the West into warring factions. But the trend of development in the Western world has switched away from—even if it has not entirely ended—the contest for external markets. To the great disarray of Communist theoreticians, the economic conflicts within the "capitalist world" are being patched up again and again, and collaboration replaces competition on an increasingly wide scale. Present attempts to set up the Common Market as a protectionist area to the exclusion of Great Britain and the United States are new water for the mills of the Communist dogmatists; but these attempts are contrary to the general trend toward integration of world economies, imposed by the scientific revolution, and can only delay but not stop this development.

The Western world moves in our time toward integration rather than fragmentation; toward the spreading of prosperity within each nation rather than the sharpening of class differences; toward more enlightened international economic policies—helping underdeveloped countries to become economically viable—rather than perpetuation of a captive-market status for colonial spheres of influence.

This trend is not uniform, and many local differences slow down or endanger it. A part of Western industrial leadership still lives in the psychology of the late nineteenth century and considers its own immediate economic advantages the only "realistic" aims of national policy. The attitude of the United States toward the Cuban revolution is a good example of such narrow-mindedness. It was influenced by financial and industrial interests which feared Cuba's transformation from a one-crop dependency of American capital into a country with a more varied and viable economy. Instead of offering the immense American economic and political power in support of this transformation, as the United States is doing elsewhere (where the primary capitalist interests are European), the United States grudgingly opposed it by economic boycott and support of attempts to dislodge the new regime. The regime was thus pushed into alliance with the Soviet Union.

What are the chances of this mistake becoming the pattern for future relations between the United States and Latin America? This future is dark and dangerous, and one or several upheavals of the Cuban type are likely to happen. One can hope, however, that the Cuban affair will prove the last error of an obsolete policy rather than the preview of future and graver mistakes. The weakness of American military intervention in Cuba was itself suggestive of halfheartedness. The time of marine landings in banana republics is over; the more constructive policies in Latin America—the good neighbor policy of the Roosevelt era and the Alliance for Progress of the Kennedy administration—are likely to be more relevant to the future. The rapid population growth in South America, which calls for a particularly rapid acceleration of production to overtake it, is a more difficult obstacle on the path of a "new deal" on a continent-wide scale than the power of American capitalist groups dreaming of puppet regimes in Latin America.

Altogether, if Western Europe and North America were left alone in the world, they could now be expected to progress slowly toward political and economic integration and increasingly widely diffused prosperity. But they are not alone.

The Main Trends in the Soviet Union

What of the trends of development in the Soviet Union? There is a lot of unnecessary discussion going on as to whether the Soviet policy is a continuation of the historic limited expansionist policies of the Russian empire or the policy of an international ideological group utilizing the resources and manpower of Russia to pursue a world-wide unlimited revolutionary program. The answer, of course, is that this policy has both roots and that their relative importance is shifting with time. The present leaders, represented by Nikita Khrushchev, are the "children of the civil war," with little or no formal education. This background of the ruling elite gives continued strength to naïve ideological commitment; but the next generation, now approaching power, which has learned ideology from books and lectures and not from personal participation in the revolution, is much more pragmatic. It is inevitable that for them this commitment will become a matter of formal belief more than of living faith, somewhat comparable to the commitment of the Western world to Christianity and political democracy rather than to the ardor of Crusaders of the Middle Ages, Moslems at the time of the siege of Constantinople, or even Hitler's SS in the early years of the Third Reich. There is in the Soviet Union no, or not much, questioning of the Communist dogma, but neither is there much willingness to sacrifice oneself, not to speak of one's whole country, on its altar.

Every ideological upheaval in history had started with a fanatical belief in the newly proclaimed truth, be it Christendom, Islam, or Communism. In every case, the fanaticism subsided with time, even if the ideological system remained formally untouched. Economic and political growth created vested interests, collective ones in a newly developed political and economic establishment and individual ones in the security and status of the members of the new ruling class and their families. The *Communist Manifesto* said that the proletariat has nothing to lose but its chains; but the Soviet leadership, from the Kremlin down to the directors of factories and chairmen of the provincial economic councils, now have a lot to lose and are disinclined to take the risk of losing it. In the real life of today's Soviet Union, not much is left of the original revolutionary mentality of absolute good in conflict with absolute evil, of a crusading revolution on the way to conquest, whatever may be written by Communist theologians in their theoretical exercises or by newspaper hacks in their editorial diatribes.

While the commitment of the Soviet Union to a world-wide Communist revolution is gradually decreasing, not so much in vociferousness as in virulence, its commitment to the traditional national values of Russia is growing. As early as before the Second World War, Stalin found it advisable to invoke Russian nationalism in support of his foreign policy. He restored the continuity of Russian history before and after the revolution in a complete reversal of Lenin's attitude after the revolution. During the war, the patriotic aspect of the conflict was stressed much more than the ideological one. It was fought above all as a "great fatherland war" for defense of Russian soil against Teuton conquerors and only secondarily as an ideological war against capitalism. The first new monument erected in Moscow after the war was dedicated to Prince Jury Dolgoruky, the founder of Muscovite power in the twelfth century!

When the war ended victoriously and large territories of Eastern Europe found themselves under Russian military occupation, the temptation to establish Communist regimes in these captive territories was too strong to be resisted. Here was a unique chance to create ideologically-bound military allies, to protect the Soviet Union from the West by a *cordon sanitaire* in reverse. Only political naïveté made some Americans hope at that time that the Soviet Union would patiently wait for peace treaties when it could create *faits accomplis,* that it would respect free elections in the occupied territories when it did not believe in them at home!

It has been repeatedly reported that Stalin looked forward to a new war with the West after the recovery of the Soviet Union from the damages of the Second World War; but it is very doubtful whether, even if Stalin were alive today, military aggression would be on the political agenda of the Soviet Union, as it was on that of Hitler's Germany. The Russian people were ready to fight and die on behalf of their fatherland in the "great patriotic war" but would not be likely to follow the leader in fighting an aggressive war for the greater glory of communism!

The question is how far, by now, has this evolution progressed? Are paroxysms of renewed fanaticism possible—a new Lenin proclaiming a complete break with the past, a second Trotsky declaring Russia a match good enough to light the fire of the world revolution and leading a Red Army onward to the West, another Dzershinsky commanding a fanatical Cheka? All that can be said is that the probability of such reverse "fluctuations" in the over-all trend of the Soviet society is small. Even in the absence of constitutional guarantees or of freedom of political debate, the mood of the nation —including the most vital elements in its Communist leadership— will act as a brake on attempts to return to extremes of terroristic rule and reckless foreign policy.

The more successful the Soviet economy, the higher the standard of living of the Soviet people, the greater will be the national concern with the preservation of these achievements and with further improvement of life conditions. *L'appétit vient en mangeant.* The American public opinion and political leadership have never made it clear to themselves whether to consider economic advancement of the Soviet Union a "good" or a "bad" thing. Of course, it is both: It is bad in that it strengthens a potential adversary biologically and physically; it is good in that it makes this adversary more contented and hopeful and less likely to risk losing his achievements in a desperate gamble. In my opinion, the second consideration far outweighs the first.

From this point of view, there is no reason for rejoicing in the obvious failure of Soviet agriculture to match the progress of industry. Of course, it is a satisfaction to Americans to see confirmed, on a continental scale, their widespread belief that productive agriculture requires personal concern of an individual farmer with the land he tills, that agriculture is best served by a strong and prosperous farmer class. In the Epilogue to *Dr. Zhivago* a spokesman for Pasternak says that everybody in the Soviet Union knows that collectivization of agriculture was a failure, but nobody dares to say it and to act accordingly. If the Soviet Union could follow Poland in reversing collectivization, much—but not all!—of its agricultural troubles would probably disappear. However, the socialization of agriculture in the Soviet Union, as contrasted with that in Poland, probably has passed the point of no return. It is most likely that the Soviet Union will not abandon collectivized agriculture, but experiment with its reorganization and increase investments in it until it works halfway satisfactorily. We in the West must hope that, whether by abandoning the collectivized system of agriculture or by making it work, the Soviet Union will get over the imbalance of its present economic situation in which 45 per cent of its population work in considerable poverty to provide the other 55 per cent with a rather monotonous diet. (In the United States, 7 per cent of the total population provides the country with an overabundance of proteins, fats, fruits, and vegetables.) This would further increase the peaceful "have" psychology of the Soviet people and make its leadership less likely to engage in reckless policies.

To sum up: The *masses* standing behind the Soviet world-political drive are large, both in numbers and in technological equipment; but the *forces* that drive them toward aggressive policies are weaker than is widely believed in America and are growing weaker with time.

Whether the trend toward the lessening of ideological fanaticism and growth of an economically viable society will be coupled in the Soviet Union with a trend toward political liberalization is less certain. By their very nature, totalitarian political systems have great difficulty in evolving toward greater spiritual and political freedom. In all likelihood, the political evolution of the Soviet state will not be smooth. As earthquakes are needed from time to time in the evolution of distorted earth layers toward a more stable equilibrium, so political earthquakes may be inevitable in the stabilization of a political system originally formed in the volcanic eruption of a social revolution, followed by rapid crystallization of the molten lava. It is, however, not to be expected that these upheavals will be directed against the fundamentals of the Soviet social and economic order, which popular masses have gradually accepted and now have no intense motivation to change. The upsets will be within the system, affecting the ruling political groups and personalities but not the basic structure of the society.

In the past, Soviet leadership has been unable to carry out any political liberalization plans because as soon as opposition was permitted to raise its head it became an opposition to the regime as such. For this reason the "Stalin Constitution," with its democratic provisions, remained (and still largely remains) a sham. Similarly, the Chinese attempt to "let a thousand flowers bloom" quickly revealed widespread dissatisfaction with the Communist regime as such and had to be abruptly terminated.

With the growing prosperity of various economic groups in the Soviet society, the path is being opened for opposition trends within the system, clamoring for an independent role in the determination of national affairs. The system of "democratic centralism," which permits dispute only within the party councils but requires absolute discipline once party leadership has spoken, has survived in the atmosphere of permanent danger to the regime as such; with this threat gone, the compulsion not to fight for one's ideas (and group interests) outside the secret party conclaves must weaken. Within the next ten or fifteen years, the Soviet leadership will have to accommodate itself to the existence of some forms of opposition and the need to create legal forms in which these groups can operate.

Europe and the Underdeveloped Continents

The main trends within the two superpowers, the United States and the U.S.S.R., point toward cautious relaxation. This eases, but does not end, the dangerous state of affairs in areas where the interests of these two superpowers clash. The crises in these areas do not create the metastability of the world situation, which follows from the very existence of large stocks of nuclear arms, but they make this metastability more dangerous, in that human failures or technological accidents can be particularly critical at the height of a political conflict, such as the world has experienced in October, 1962, over Cuba. How far the "background" of the U.S.-U.S.S.R. relations will be kept from the blowup point depends very largely on the policies of the two superpowers. *De facto,* both sides have entered, since Cuba, into a cooling-down period; but so far, no deliberate decision has been made to stabilize it and to put all local conflicts into a "deep freeze." This decision is mostly up to the United States because, at least as far as Europe is concerned, the West is the "revisionist" (the Soviet speakers like to call it "revanchist") side, whereas the Soviet Union is the satiated and therefore conservative side. The stabilization of Western European economies since the Marshall Plan has put an end to plans of the Soviet Union to revise further in its own favor the political map of Europe; since that time, Soviet aspirations have become concentrated on holding on to what has been acquired. The exposed position of West Berlin is being used by the Soviet Union to exert pressure on the West for formal recognition of its acquisitions—of Soviet annexations in the Baltic and East Prussia, of the Oder-Neisse border between Poland and Germany, and of a second, Communist German state. The United States could have decided on such recognition soon after the war; and it might have been the wise thing to do. But the Acheson-Dulles policy was to keep the situation in Eastern Europe as unsettled and uncomfortable for the Soviet Union as possible, in the expectation of a political upheaval there, while at the same time refusing American military support, without which such upheavals had to go the way of the East German uprising of 1953 and the Hungarian revolution of 1956.

In recent years, greater dependence of the United States on its Western European allies, by now restored to their full economic, if not military, power, makes it increasingly difficult for the United

States to do what it may now acknowledge as advisable—to recognize the *de facto* frontiers and regimes in Eastern Europe. It is therefore to be expected that instead of decisive abandonment of all support for revisionist aspirations in Europe, the United States will continue to extend only grudgingly, step by step, its *de facto* recognition of the status quo, thus keeping the political temperature in Europe dangerously hot. This continued tension will become increasingly dangerous with time. The present asymmetry of nuclear threats from the East and the West, which led to the American victory in the Cuban conflict, is bound to disappear through increase in the number of American and Soviet hardened missile bases at home and invulnerable rocket launchers on submarines, devaluing the present advantages to America of its advanced European and Asian bases and of its now unique Polaris-armed underwater fleet. With the threat of a first nuclear strike becoming equally plausible (or rather, equally implausible!) on both sides, crises of brinkmanship could be increasingly risky, because both sides will become as unwilling to retire from the brink as the United States was in the Cuban conflict. The game of "chicken" may thus go on longer, and the danger of an "accidental" war increase in proportion.

Will internal developments in the countries of western and eastern Europe permit the stabilization of the present order of things in Europe? In western Europe, critical upsets are unlikely. Upheavals such as a Communist election victory in Italy, or a new nationalist wave in West Germany, or a popular front revolution in France, are nowhere in sight, and progressive economic and political integration, increasing the viability of Western Europe, is indicated. The greatest single obstacle to this integration—De Gaulle's belief in France's mission of leading Europe, with, or preferably without, Great Britain, and hopefully including Russia, toward conversion into a third world power beside the United States and China—is not rooted in common feelings of the European masses and is likely to disappear with the end of his personal rule, be it in five or ten years. The delay in Britain's joining the Common Market is prolonged by this attitude, and this may cause a revulsion of British opinion against it, but ultimately, Britain has nowhere else to go. That England "does not belong to Europe" was, and still is, a British conception; Europeans have always been bewildered and, to some extent, amused by Englishmen asserting that the British Isles are not a part of Europe. If in Britain itself the outward pull of the Empire in its much attenuated "Commonwealth" form fails to sustain the traditional insular—or, rather, oceanic—psychology, European resistance to British association, based, as it is in De Gaulle's case, on personal mystique and irrational resentment more than on rational considerations, will not be prolonged or effective. British association with Europe is one of the inevitable things which can be long delayed under peacetime conditions but which are bound to come sooner or later.

Stabilization and progressive integration are much less certain in Eastern Europe. In this area, two trends are in conflict. One is the trend, parallel to that in Soviet Russia itself, toward the acceptance of the now existing social and economic order, of permanent association with the Soviet Union in a Communist commonwealth, and of gradual evolution toward greater independence, and a certain internal liberalization, within this association. This way is beset with difficulties. In the countries of Eastern Europe, opposition to the ruling party groups is still apt to explode, as it did in Hungary, into opposition to the system as such and therefore cannot be tolerated by the Communist leadership.

It is logical to presume that, in the long run, the countries of Eastern Europe will free themselves from their present subservience to Moscow; but the possibility of this liberation occurring by peaceful changes within the system, not provoking Soviet military intervention, is problematical. In some countries, such as Bulgaria—perhaps, also in Czechoslovakia—which have had only a limited tradition of national independence and a long tradition of pro-Russian attitudes, the development may follow that in the Soviet Union itself, with the economic system becoming increasingly acceptable to the people and a certain liberalization occurring without threat to the survival of the system. Chances for such a transformation are much smaller in nations with strong nationalist and anti-Russian traditions, such as Poland and Hungary.

Poland, in particular, is tied to the Soviet commonwealth primarily by the support it needs against German revisionism. Western recognition of the Oder-Neisse border and decisive opposition to any German plans to upset it could weaken this tie dangerously and put an end to the Polish commitment to support the East German regime. Poland, similarly to Yugoslavia, could then regain considerable freedom of action. The Soviet Union, in continuation of its present trend, may not oppose this process but accommodate itself to it as long as it does not upset the strategic balance in Europe. (The fateful error of the Hungarian revolutionary regime was to announce its quitting the Warsaw military pact.)

The greatest Communist headache in Europe is, of course, East Germany. Lavrenti Beria, Stalin's last terrorist-in-chief, recognized the difficulty of keeping one-third of the German nation tied up in a Communist dependency and advocated its release from Russian grip. Beria's downfall, which followed the Berlin uprising of June, 1953, closed this avenue of retreat. The Soviet Union drifted into the position of the proverbial bear hunter who had caught a bear but could not bring it home because the bear did not want to go, and could not come home himself because the bear did not let him go. There is very little chance that the D.D.R. (Deutsche Demokratische Republik) will become a viable German state, standing on its own legs and accepted by its population (as independent Austria has become accepted after the experience of the *Anschluss* under Hitler). It is more likely to remain indefinitely a bleeding sore in the body of the Communist alliance, a "sick man" of Eastern Europe. The "operation Berlin" has for the Soviet Union the one purpose of propping up, by foreign recognition, the shaky D.D.R. Gradually, step by step, this operation is likely to succeed, because the West is unable to support the revisionist aims of West Germany by force and will have to accommodate itself grudgingly to the *de facto* existence of the D.D.R. as the price for continuation of the independent status of West Berlin. However, even full diplomatic recognition of the D.D.R. would provide only a momentary shot-in-the-arm for this Communist rump state and would increase only temporarily the malleability of its population. One has merely to imagine the context in which the problem of migration from East into West Germany will appear after the recognition of the D.D.R. and the consequent loosening of its border controls. The best policy for the West would be to press for the implementation of the Soviet proposals for a confederation between the two Germanys. The D.D.R. would not be able to survive such a confederation, which would require, at the very least, freedom of travel, newspaper and book circulation, and other exchanges between the two confederated states. The confederation program, like the program of "complete and universal disarmament," should have been the fighting program of the libertarian West and not of the totalitarian East!

The continued existence of unstable Communist states in Eastern Europe, particularly of the D.D.R., is bound to delay political and economic stabilization and relaxation of tensions in Europe, and repeated fluctuations in the area will bring the West and Soviet Union into "brinkmanship" situations. However, since neither of the two sides will ever be in the position of deciding "to get it over with" by the test of arms, these fluctuations will end in the same way that almost all fluctuations end in nature—a return to *status quo ante*. At least, that is what one may expect under the assumption underlying this analysis, that the major trends of events will not be overturned by accidents or irrational acts of small groups.

Since future displays of brinkmanship will not lead to anything except periods of dangerous instability, it would be a wise policy for both sides to renounce all political gambits likely to lead to such situations and, instead, to collaborate in the stabilization of the status quo, either openly or—as long as this proves impossible for reasons of domestic policies—at least, *de facto*. How much, if any, of this political wisdom will be shown by the West or by Moscow one cannot predict; but the ultimate result remains the same: The only alternative to the catastrophe of an unwanted nuclear war is preservation of the existing political map of Europe, and the only change the West can reasonably expect there is slow evolution of the countries of eastern Europe to a more independent and open status. It is conceivable that, at a certain stage of this evolution, Moscow will be willing to let the D.D.R. go as a heavy drag on its own security, perhaps by pushing it into a confederation with Western Germany, despite the knowledge that such a confederation is likely to go the way of the German confederation of 1866—becoming the first step toward a unified German state dominated by its strongest component, Prussia then and the Federal Republic of Germany now.

Underdeveloped Countries

Are there any other areas of conflict which may endanger the process of gradual accommodation between the United States and the U.S.S.R.? It is most likely that, whether the Cuban communist state survives or not, other areas in South America—Venezuela, Peru, Brazil, or some smaller Latin American state—will experience, in the next two decades, violent social upheavals. I believe, however, that the Soviet Union, while sympathetic to these upheavals, will stay out of direct intervention and that the United States will do the same, so that these revolutions will be given the chance to find their own way. The result is likely to be a radical but pragmatic reform, a national rather than a Soviet-type brand of socialism. The United States could, of course, oppose these revolutions by ineffective interventions, similar to the one attempted in Cuba, and thus push the revolutionary regimes into the arms of the Soviet Union; this seems "against the trend" of American policies, but great American private investments in South America may exert a powerful backward pressure on this trend. Therefore, events in Latin America are the most likely source of critically dangerous world situations in the next quarter-century.

The situation is not so dangerous in Africa, because American vested interests are much smaller there and the possibility of such interests' upsetting a rational and enlightened policy is less likely. It is much easier for the United States to act rationally when the threatened investments are English or Belgian and not American!

In Asia, the opposing forces of the West and the Soviet Union stand already now in the shadow of China. The conflict between the two present superpowers over, say, Iraq, Iran, or Turkey could have led to war between 1945 and 1960; but, luckily, it did not. In the future, both sides will have an increased interest in keeping this part of the world stable to avoid provoking Chinese interference. The danger of Chinese intervention has permitted the patching-up of the East-West conflict even in remote Laos, on the very doorstep of China. Disregarding Chinese intervention is still possible now, but may become impossible ten or twenty years hence, anywhere in Asia.

To sum up, the big masses of the Soviet Union and the United States are not now on a violent collision course in any of the conflict areas in Europe or Asia and may enter into collision only as a result of an unexpectedly reckless policy on one or the other side. The most dangerous potential source of such collision is Latin America. It is in the mutual interest of the United States and the U.S.S.R. to convert their present *de facto* community of interests in general stability into a common, established policy, to cool down the hot climate of their relations, and to reduce the danger of a small fluctuation's destroying the precarious peace between them.

CHINA

China is the great puzzle of the future. Two major trends are obviously at work there: the Messianic drive of a young Communist movement whose leaders think that the mantle of revolutionary world leadership has descended on them and the pressure of an already enormous and rapidly growing population. Together they create a potent aggressive force, acting on an immense mass of people.

The Chinese revolution is obviously "out of phase" with the Soviet revolution. Whether it will go through the same cycle, and ultimately lead to the same type of stable society, we do not know; but at present, the difference between these two Communist empires is getting wider, as the Soviet Union settles into a relatively prosperous and relaxed existence, while China goes through its leanest years of hunger and destitution, stimulating the revolutionary fervor and permitting no relaxation of iron discipline within its leading elite.

This phase difference puts a great strain on Chino-Soviet relations, exacerbated by the patronizing attitude of the more "mature" Russians and the racial pride of the Chinese, conscious of representing a much longer cultural tradition than the white nations, including Russia, and long smarting under semicolonial treatment by the latter. The bond of ideological brotherhood between the two Communist empires is obviously close to snapping, and each of the two is ready to go its own way. The separation could be gradual, repeatedly masked by the cloak of a continuing formal alliance concealing the truth of political estrangement. The Chinese could remain satisfied with the acceptance of their leadership by the Communist parties in southeast Asia and not interfere with Soviet leadership of the Communist movements in the rest of the world. However, already the case of Albania suggests that such tacit division of Communist spheres of influence is not likely. "Internal Chinese" exist in many Communist parties in Europe and elsewhere, and Chinese leadership is unlikely to renounce exploiting them to extend its influence. The Soviet leadership cannot let this happen, if for no other reason than that its own "Chinese" faction is not negligible. It is therefore most likely that in the next few years—perhaps after additional attempts to patch up the differences by new international gatherings, such as the one that was held in Moscow in November, 1960—the Communist movement will openly split, and a Chinese-led Fourth (or Fifth?) International will follow the Moscow-sponsored Third. Who is likely to follow Peiping in this break?

It is rather safe to predict that the Communist movements in Western Europe will stay with Moscow. The laboring classes in Europe are growing increasingly prosperous and conservative, and this imposes an increasingly cautious and reformistic policy on European Communist parties, particularly mass-membership parties, such as those in Italy, France, or Finland. The "Chinese" faction is stronger in the Soviet Union and in some Eastern European countries, where life is much harder and Communist dogmatism therefore finds more numerous adherents (although apparently more among intellectuals, such as the theoreticians from the Marx-Engels Institute in Moscow, than among labor unions and practical administrators). Acceptance of Chinese slogans of inevitable violent world revolution is also unlikely—at least, in the foreseeable future—in the new countries of Africa, where the revolutionary spirit is channeled primarily into nationalism and where truly proletarized population masses, susceptible to anticapitalist fanaticism, do not exist. The situation may be different in South America, or at least in certain parts of it, where a genuinely revolutionary leadership can find more susceptible human material. These revolutionary masses consist of impoverished peasantry rather than of industrial proletariat; but after all, similar disgruntled peasant masses brought about the success of the Bolshevik revolution in Russia in 1917. Disappointed in Soviet leadership since its retreat from Cuba, these masses may be ready to follow the Chinese banners.

Chinese Communist society is still in the incubation period, similar to the 1920–33 period in Soviet history. After the failure of the first overoptimistic Soviet attempts to carry the Communist revolution westward on the bayonet tips of Soviet soldiers and to organize Comintern-sponsored armed uprisings, such as the Max Hölz insurrection in central Germany in 1921, the Soviet revolution withdrew into itself to "build communism in one country." It reappeared on the world stage twenty-five years later, in the Second World War, in possession of an enormous military establishment and a vast production apparatus but minus much of its original revolutionary *élan.*

When Mao Tse-tung and Khrushchev fight each other with quotations from Lenin, the Chinese leader has a more convincing case—the present Chinese policy is closer to Soviet policy in the early period of the revolution than is the Soviet policy since Stalin's death. But, as Marxism preaches, *Das Sein bestimmt das Bewusstsein;* and the reality of existence of the Soviet Union has changed radically since the early twenties. Is the same change of *Sein,* and thus also of *Bewusstsein,* in store for China?

The failure of the "great leap forward" had demonstrated—what should have been obvious from the beginning—that China could not compress twenty years of Soviet development into five years of its own and thus combine large industrial power with intact revolutionary fervor. Considering the much weaker original technological and scientific level of prerevolutionary China compared with prerevolutionary Russia, it is to be expected that it will take China more time than it did Russia to become an advanced industrial power. If Russian physicists could catch up with the wartime development of nuclear arms in the United States (which itself took about five years) in about six years, a similar achievement should take China ten. (The first Chinese atom bombs would then make their appearance not before 1965.) But however long it may take, one must prudently assume that within two or three decades, the Chinese leadership, if it avoids premature military showdowns (such as it seemed for a while to seek in India), will emerge in possession of a military and industrial power of the same order of magnitude as that of the other major world powers.

The Korean War and the brief Himalayan campaign made it clear that the Chinese Communists have already succeeded in building up an effective military machine. This is natural; even in democratic countries, military machines are hierarchical organizations, based on absolute obedience and elementary indoctrination—mechanisms which a totalitarian regime is quite competent to create and handle. Trotsky, and the tsarist generals mobilized by him, built up an effective Red Army in Russia at the time when administration, industry, transportation, agriculture, education, and science were still in a state of collapse and anarchy.

Heavy industry was the next area that proved, in Russia, to be susceptible of rapid development under centralized leadership. The area of economic activity that has shown itself, in all Eastern Europe, most refractory to centralized leadership is agriculture, because it is still largely a primitive art, requiring much individual instinct, insight, and devotion.

Here, a parting of ways seems to announce itself between Communist China and Communist Russia. The Soviet Union had let agriculture mark time—if not go to the dogs—while frenziedly building up heavy industry. To provide the army and the industrial workers with a minimum of food, the regime resorted to mass terror in the villages, to the "extermination of the kulaks as a class." (All prosperous and independent peasants were relegated to this class.) The Chinese, after the failure of an even more ambitious attempt to make a "great leap forward" in industrialization and an even more frenzied establishment of land communes, made a turnabout and decided to give highest priority to the development of agriculture, letting heavy industry "mark time." Two results are conceivable: The first is that this deviation from the royal road to communism through development of heavy industry will prove as short-lived as has been the NEP (New Economic Policy) proclaimed by Lenin in 1921. The other alternative is for the concentration on agriculture and consumer industries to become serious and long-lasting. In this second case Chinese communism may become bogged down and perhaps even gradually transform itself into the "agrarian reformism" for which some Americans have at times taken it. This course would make the development of Communist China quite different from that of Soviet Russia, a development which could be considered in better agreement with the greater predominance of peasantry in China, even in comparison with the also predominantly agricultural prerevolutionary Russia; this course would also be much more desirable from the Western point of view. However, the first alternative appears more likely. Continuing preoccupation with their own agricultural problems would make it practically impossible for Chinese Communists to sustain their drive for world-wide revolutionary leadership, even in countries such as Brazil, where land-hungry peasants constitute the main revolutionary material. Therefore, it is more likely that after a pause of a few years, Chinese leadership will again concentrate on heavy industrialization. If this is the case, Communist China may emerge onto the arena of "big politics" in twenty or thirty years in possession of a powerful military establishment, including nuclear weapons, and a vast industrial machine but with agricultural production still lagging behind. How much of its revolutionary *élan* will still be alive we do not know; but its population pressure is bound to remain and grow.

There are many who predict that Chinese population pressure will exert itself above all in the direction of the more thinly populated lands beyond the Chinese frontiers to the northwest—Mongolia, Siberia, and Russian Turkestan—and that this land hunger may even lead to military aggression, as it did in the case of Japan. One can even envisage an alliance between revolutionary China and Japan in pushing the white man out of northeast Asia to give more living space for the yellow man. The present Japanese prosperity, after the loss of all its overseas possessions and captive markets, is a dangerously unstable "miracle" and likely to be replaced, sooner or later, by a renewed struggle for greater lebensraum. Soviet political experts must worry about this possibility. However, one has to consider that by the time China will be at all capable of active power politics, the nuclear armaments of the Soviet Union will be even stronger than they are now. It is easy to say—and still easier to repeat—that China, with its now six hundred million, and by then close to a billion, inhabitants, will be able to risk, and survive, the loss of several hundred millions of them in a nuclear war; but this is plain nonsense. China's economy is as vulnerable to the kind of nuclear attack made possible by thermonuclear weapons as that of any other country; and with growing industrialization, this vulnerability will increase. Thermonuclear weapons are even now capable of covering all China with deadly radioactive fallout and inflicting intolerable damage to its economic viability. An expansionist Chinese drive will have to seek outlets in directions in which no certainty of powerful nuclear opposition may exist. This could be the case in southeast Asia (Vietnam, Thailand, Philippines, Indonesia, Australia), central Asia (India, Pakistan), or southwest

Asia (Iran, Iraq)—unless expansion in this direction, too, is blocked by certainty of nuclear reaction.

It was suggested at the beginning of this paper that the time scale of possible happenings may well determine which of them will have the chance of happening at all. The question whether China's population pressure will permit it to expand toward the south and southwest depends on whether, by the time China becomes strong enough, militarily and technologically, to exert this pressure, the West and the Soviet Union will have arrived at a general accommodation. In the face of the combined opposition of the West and the U.S.S.R., Chinese expansion into southern Asia will certainly be impossible; but if the nuclear power of the West is neutralized at that time by Russian nuclear power, such expansion may become possible. China could foster revolutionary movements in certain Asian countries—in Burma or Malaya, in Iran or Pakistan—and the West might be unable to oppose these movements militarily (as it is doing now in Vietnam) because of the Soviet threat of intervention.

The two trends—the drawing together of the two present major "have" powers, the United States (and Western Europe) and the U.S.S.R., and the growth of the revolutionary "have-not" power, China—run in competition. If, as I believe more likely, the United States-U.S.S.R. accommodation is achieved first, spurred as it is by the common apprehension of China, then Chinese expansionism, despite its twin moving forces of population pressure and (at least, at the present time) still virulent revolutionary fervor, will be forced to mark time, until Communist China, too, will have gone the way of other revolutionary movements, settling into some kind of a stable society, with a new ruling class and an overriding interest in its own preservation. But the leadership of the United States and the U.S.S.R. must hurry to comprehend the challenge and act in time to meet it!

Disarmament

One may wonder why the word "disarmament" has not yet occurred in this presentation. Is not the arms race generally supposed to be the greatest threat to peace? Is not the danger of war by misunderstanding or technical accident the consequence of the arms race; and is not disarmament—in particular, nuclear disarmament—the only way to eliminate this danger? Is not disarmament the first area in which the United States and the U.S.S.R. have a common interest and a chance of mutual accommodation?

This is logical, and yet it disregards the impossibility of separating military from political programs. As long as nations pursue conflicting power policies, they will not be able to end competitive armament drives. The current wide awareness of the enormous costs and deadly dangers of the nuclear arms race, which is common to all peoples and has communicated itself to all governments, has led to almost continuous disarmament negotiations. These have gone on, with brief interruptions, for the last fifteen years, but without success. The awareness of common danger could not change the stark but simple fact that disarmament remained—and still remains—impossible without previous (or at least simultaneous) political settlement. The disarmed state, as well as the state of "stabilized nuclear deterrence" or "arms control" aimed at by some American theoreticians, is a state of equal military status of both sides; this equality cannot be accepted by the two competitors except in combination with an equality of political status. This the West refuses to give to Soviet Russia and will continue refusing as long as it considers the Soviet political aim to be the West's destruction.

Certain partial agreements, such as that on cessation of nuclear tests or on over-all limitation of armed forces, can and probably will be reached and may produce dangerous euphoria—"Hallelujah, peace is come"; but such partial successes will not change the fundamental fact of our time—that the major nations possess the means for sudden and complete destruction of all other nations. Such agreements will not end—although they may moderate—the critical metastability of the world situation.

The only possibility of a substantial slowdown, or complete elimination, of the arms race lies in a world-wide political settlement; political warfare must be ended first before the arms race can end. As between the United States and the U.S.S.R., common interests in stabilization of the political setup in the world are so strong, compared with their interests in winning more power at the expense of each other, that *in abstracto* such a world-wide settlement appears possible. However, the recognition of this community of interests of the two successful superpowers of our time is as yet dim and in many leading groups altogether non-existent. The approach to *de facto* stabilization is therefore likely to be, particularly on the American side, tentative and grudging rather than radical and wholehearted. Under these conditions, far-reaching disarmament agreements are not in the cards for the next two decades. What the world can hope for is that after a decade of *de facto* absence of acute political crises between the U.S.S.R. and the West, and with increased East-West co-operation in various technical and economic areas, the endless, frustrating, all-or-nothing disarmament negotiations will yield place to a succession of partial agreements, ultimately ending in a significant reduction of armaments. However, even this delayed and gradual progress may be by then endangered by the great question mark of the growing military might of China.

Revolutions Ahead?

Another sceptic may ask: And what about new revolutionary forces that may arise in the next ten or twenty years? There is so much conscious and subconscious expectation alive in the West that the "unnatural," "godless," "tyrannical" Communist systems are going to be upset one day by the uprising of their own masses, disappointed in their yearnings for political freedom and economic prosperity! And, symmetrically, so much conscious and subconscious expectation survives in the Communist states that one day laboring masses of the West will rise against the iniquities of the capitalist system. It appears, however, that both expectations are unreasonable, at least as far as the next quarter of a century is concerned. No political or economic system collapses unless it first shows itself incapable of ruling or producing effectively and unable to maintain the active support of an energetic minority and tacit acceptance by the mass of the population. Neither the libertarian system in the United States and Western Europe nor the Communist system in the Soviet Union shows signs of such internal decay. Both operate effectively in the technological and economic areas; whether the production growth is at the rate of 3, 5, or 8 per cent annually is not important as long as it is growth. Both are able to maintain the enthusiastic support of the most active minority and the acceptance of the politically inactive majority. Both seem to show the characteristic which is as decisive for a viable society as for a viable individual organism—the capacity of adjusting itself to cope with new conditions of existence.

Since the new conditions of life are imposed by the scientific revolution and are therefore largely the same on both sides of the iron curtain, these adaptive changes are largely parallel and increase the similarities rather than accentuate the contrasts between the two societies. I do not suggest, as do some analysts, that the United States and the U.S.S.R. are bound to become more and more similar; the difference in the hierarchy of societal values remains fundamental and cannot be erased even by twenty years of technological revolution. The West is committed to a value hierarchy which places individual achievement, individual freedom, and individual satisfaction above service of an individual to society; the Soviet Union is committed to a reverse order of priorities. Both value systems seem to be able to support a viable, growing society, although certain efforts prosper better in one type of society and others in the other type. To those who may suggest that the difference is small, I would answer with the famous remark, "Vive la différence," with which a French deputy answered the arguments of a violent feminist minimizing the differences between the two sexes. The individualism of free society is the most important inspiration of Western achievements for humanity; society-devotion is, similarly, the main source of achievements in the Soviet society. We may dream of a synthesis of the two; but as Goethe's Epimetheus said: "Des tätigen Manns Behagen sei Parteilichkeit"—creative efforts are most effective when based not on contemplative synthesis but on partisanship for one idea, even if it means blindness to others.

Revolutionary movements, and changes caused by them, are not unlikely in the next twenty years, particularly in some countries of Eastern Europe; but there will probably be changes within the existing society and not revolutions against this society.

What about political changes imposed by new scientific and technological upheavals? Truly revolutionary scientific events are unpredictable, much more so than supposedly unpredictable political events, because, similar to great achievements in art, they are creative breakthroughs of the human mind. However, with some degree of probability, one may predict that the most revolutionary developments in science of importance to society are likely to take place in the next few decades in biology. New discoveries in physics may give us understanding of the world of elementary particles, bringing order into chaos, as quantum theory brought order into the world of atoms and molecules; it is even possible that they will give access to new, more powerful elemental forces than those mankind has obtained by its first penetration into the atomic nucleus. Discoveries in astrophysics, made possible by manned and unmanned space flights, will add largely and unpredictably to man's understanding of the universe but will not change significantly the material life on earth or give new economic or military assets to the nations achieving them.

The progress in biology is more likely to confront mankind with great new practical challenges. The mastery of heredity, free decision over the sex of children, capacity to influence the mental makeup and the psychological attitudes of humans, the possibility of prolonging significantly the life span of individuals, the elimination of detrimental genes from the hereditary plasm, artificial creation of living matter—these are among the problems toward the solution of which modern biology is advancing with a speed that may any day bring about an unpredictable breakthrough. Many of these solutions will confront society with heavy new responsibilities. Who is to decide on the application of new fateful knowledge; what general principles should be used in the application of new discoveries? For example, if methods are found to prolong life to, say, one hundred fifty or two hundred years, whose lives should be prolonged? Should making space for children be given preference to prolonging the life of oldsters, or the other way around? If human population on earth is to be stabilized on some optimum level, who is to decide where that level lies? These will be fateful questions confronting all societies—democratic or totalitarian, libertarian or communistic. Even in the military field, biological discoveries may change the situation more significantly than new nuclear weapons or new means for their delivery.

It is unlikely, however, that any one of these fateful challenges will become acute in the time span to which this essay applies—the next twenty-five years. If this proves wrong, the effect may be similar to that which an earlier perfection of the atom bomb would have had on the validity of the 1939 predictions about the course of the Second World War!

What To Do about It All?

If these estimates of trends now at work in the world political stage are correct, what do they mean for American political leadership? Two choices offer themselves: to follow the existing trend, to sail before the wind, or to stem it.

One thing is certain—a powerful trend cannot be stemmed by weak opposition. If the United States wants to upset the existing trend, this can be done, not through halfhearted measures, but by full application of American power, directed by a strong will, as General de Gaulle now opposes the trend toward increasing political and economic integration of the West. This is what is wrong with much well-meaning advice given to American policymakers. A small first step in a certain direction is often advocated without reasonable hope that it will be followed by stronger further steps, whatever the result of the first one and however strong the opposition to it is likely to be. Thus, people who suggest a policy of unilateral disarmament steps, or of political concessions, do not consider that if the United States government takes their advice and tentatively engages on the suggested road, it is certain to meet, at first, with political defeats—long before the revolutionary effect of a consequent policy of the suggested type could appear—and that strong national and international pressure will soon reverse the policy unless it is undertaken with an iron determination and readiness, from the beginning, to go to the end, not to be detracted by reverses, and to buck all storms.

The same applies to "activist" steps, such as the overthrow by force of Castro's regime in Cuba. Unless the United States has the will not only to face the waves of hostility beating in its face from all over the world after such a display of force but to follow it by other applications of force, to install and maintain pro-American regimes in every country threatened by social revolution, such a policy will be an ultimate failure, because nothing short of world-wide victory over the forces of social revolution will vindicate it. In hot or cold war, pinpricks or local victories are of no ultimate avail; you merely strengthen the regime of a Franco, a Castro, or a Mao Tse-tung by non-recognition, sabotage, quarantine, or refusal to trade. In the same way, a policy aiming at the establishment of a world government, or another new international authority, cannot lead to success unless it is entered into not tentatively but with full strength and is pursued with persistence and a national determination which will be able to survive inevitable initial disappointments, detriment to America's international power, and damage to its alliances.

The alternative of following the trends of the time has the advantage that individual steps, however small, if they point in the "right" direction are not likely to be reversed, as is likely to happen to small steps against these trends. In the present situation, the trends suggested above are not unfavorable to United States interests or incompatible with American ideals. One can imagine a highly imaginative and revolutionary American policy concordant with the direction of these trends, but such policies would require, to begin with, a revolutionary change in American public opinion, of which there is at present but few signs. What the American government could do, however, is to sail before the wind of time with a stronger will and a firmer hand at the tiller than it has now. Only in this way can we shorten the deadly dangerous time during which the sword of a nuclear war "which nobody wants" will be hanging over mankind and create, as rapidly as possible, a stable state of East-West relations, not easily destroyed by accidental "fluctuations" or by ill-will of a few men.

ADDENDUM TO PART I

A Report to the Secretary of War June, 1945

INTRODUCTION

The following report was submitted to Secretary of War Henry L. Stimson by a committee "On Social and Political Implications [of Atomic Energy]," appointed in the spring of 1945 by the director of the "Metallurgical Laboratory" (code name for the Atomic Bomb Research Laboratory at the University of Chicago), Farrington Daniels. An Interim Committee had been appointed by the Secretary of War to consider the postwar future of atomic energy and to solicit comments from the several laboratories of the Manhattan project. Scientists at the Metallurgical Laboratory, deeply concerned with the impact that the atom bomb, about to be perfected, was bound to have on the future of mankind, seized this opportunity to acquaint the government with their concerns.

The committee consisted of four physicists, James Franck, the chairman (after whom the report became known as the "Franck Report"), Leo Szilard, J. C. Stearns, and Don Hughes; one biologist, J. J. Nickson; and two chemists, Glen Seaborg and myself. The report was secret; but in 1946, the classification was removed, and it was published in the *Bulletin of the Atomic Scientists* in May, 1946.

I permit myself to reprint the Franck Report here because of my participation in its preparation and because it best represents our collective thinking at that critical time. Most of the report was written by me; but the emphasis on the use (or rather, non-use) of the bomb in Japan, which has given the report its main historic significance, was due to James Franck and Leo Szilard. This matter was much on their minds because of their greater awareness of the impending decision about this use by the leaders of the Manhattan project, whereas my own attitude was more that of an outsider concerned with general implications of the release of nuclear energy. To some extent, this difference still persists; the concern of many of my colleagues remains primarily with the immediate political decisions, whereas my interest is still centered on the long-range implications of the scientific revolution, of which the creation of atomic weapons was only one, even if the most striking, event to date.

The Franck Report was lost somewhere in the higher echelons in Washington. Only vague references to it are found in the memoirs of the then secretary of war, Henry L. Stimson, and other contemporary documents. The decision to use the two bombs, available after the Alamogordo test, on two large Japanese cities was supported by the top scientific leaders of the Manhattan project. In addition to the decisive consideration of saving the many lives that would be lost in an invasion of Japan, two other considerations seem to have played a role: the need to demonstrate to Congress that two billion dollars had not been spent in vain; and the reluctance of the scientific leaders of the project to guarantee that a demonstration of the bomb before foreign observers would not be a dud, in which case the effect of the demonstration might have been to stiffen Japanese resistance.

THE FRANCK REPORT

PREAMBLE

The only reason to treat nuclear power differently from all the other developments in the field of physics is the possibility of its use as a means of political pressure in peace and sudden destruction in war. All present plans for the organization of research, scientific and industrial development, and publication in the field of nucleonics are conditioned by the political and military climate in which one expects those plans to be carried out. Therefore, in making suggestions for the postwar organization of nucleonics, a discussion of political problems cannot be avoided. The scientists on this project do not presume to speak authoritatively on problems of national and international policy. However, we found ourselves, by the force of events during the last five years, in the position of a small group of citizens cognizant of a grave danger for the safety of this country, as well as for the future of all other nations, of which the rest of mankind is unaware. We therefore feel it our duty to urge that the political problems, arising from the mastering of nuclear power, be recognized in all their gravity and that appropriate steps be taken for their study and the preparation of necessary decisions. We hope that the creation by the Secretary of War of the Committee to deal with all aspects of nucleonics indicates that these implications have been recognized by the government. We believe that our acquaintance with the scientific elements of the situation and prolonged preoccupation with its world-wide political implications impose on us the obligation to offer to the Committee some suggestions as to the possible solution of these grave problems.

Scientists have often before been accused of providing new weapons for the mutual destruction of nations, instead of improving their well-being. It is undoubtedly true that the discovery of flying, for example, has so far brought much more misery than enjoyment and profit to humanity. However, in the past, scientists could disclaim direct responsibility for the use to which mankind had put their disinterested discoveries. We feel compelled to take a more active stand now because the success which we have achieved in the development of nuclear power is fraught with infinitely greater dangers than were all the inventions of the past. All of us, familiar with the present state of nucleonics, live with the vision before our eyes of sudden destruction visited on our own country, of a Pearl Harbor disaster repeated in thousand-fold magnification in every one of our major cities.

In the past, science has also often been able to provide new methods of protection against new weapons of aggression it made possible; but it cannot promise such efficient protection against the destructive use of nuclear power. This protection can come only from the political organization of the world. Among all the arguments calling for an efficient international organization for peace, the existence of nuclear weapons is the most compelling one. In the absence of an international authority which would make all resort to force in international conflicts impossible, nations could still be diverted from a path which must lead to total mutual destruction by a specific international agreement barring a nuclear armaments race.

Prospects of an Armaments Race

It could be suggested that the danger of destruction by nuclear weapons can be avoided—at least as far as this country is concerned—either by keeping our discoveries secret for an indefinite time or else by developing our nuclear armaments at such a pace that no other nations would think of attacking us from fear of overwhelming retaliation.

The answer to the first suggestion is that although we undoubtedly are at present ahead of the rest of the world in this field, the fundamental facts of nuclear power are a subject of common knowledge. British scientists know as much as we do about the basic wartime progress of nucleonics—if not of the specific processes used in our engineering developments—and the role which French nuclear physicists have played in the prewar development of this field, plus their occasional contact with our projects, will enable them to catch up rapidly, at least as far as basic scientific discoveries are concerned. German scientists, in whose discoveries the whole development of this field originated, apparently did not develop it during the war to the same extent to which this has been done in America; but to the last day of the European war, we were living in constant apprehension as to their possible achievements. The certainty that German scientists were working on this weapon, and that their government would certainly have no scruples against using it when available, was the main motivation of the initiative which American scientists took in urging the development of nuclear power for military purposes on a large scale in this country. In Russia, too, the basic facts and implications of nuclear power were well understood in 1940, and the experience of Russian scientists in nuclear research is entirely sufficient to enable them to retrace our steps within a few years, even if we should make every attempt to conceal them. Even if we can retain our leadership in basic knowledge of nucleonics, for a certain time, by maintaining secrecy as to all results achieved on this and associated projects, it would be foolish to hope that this can protect us for more than a few years.

It may be asked whether we cannot prevent the development of military nucleonics in other countries by a monopoly on the raw materials of nuclear power. The answer is that even though the largest now known deposits of uranium ores are under the control of powers which belong to the "Western" group (Canada, Belgium, and British India), the old deposits in Czechoslovakia are outside this sphere. Russia is known to be mining radium on its own territory; and even if we do not know the size of the deposits discovered so far in the U.S.S.R., the probability that no large reserves of uranium will be found in a country which covers one-fifth of the land area of the earth (and whose sphere of influence takes in additional territory) is too small to serve as a basis for security. Thus, we cannot hope to avoid a nuclear armaments race either by keeping secret from the competing nations the basic scientific facts of nuclear power or by cornering the raw materials required for such a race.

We shall now consider the second of the two suggestions made at the beginning of this section and ask whether we could not feel ourselves safe in a race of nuclear armaments by virtue of our greater industrial potential, including greater diffusion of scientific and technical knowledge, greater volume and efficiency of our skilled labor corps, and greater experience of our management—all the factors whose importance has been so strikingly demonstrated in the conversion of this country into an arsenal of the allied nations in the present war. The answer is that all that these advantages can give us is the accumulation of a larger number of bigger and better atomic bombs.

However, such a quantitative advantage in reserves of bottled destructive power will not make us safe from sudden attack. Just because a potential enemy will be afraid of being "outnumbered and outgunned," the temptation for him may be overwhelming to attempt a sudden unprovoked blow—particularly if he should suspect us of harboring aggressive intentions against his security or his sphere of influence. In no other type of warfare does the advantage lie so heavily with the aggressor. He can place his "infernal machines" in advance in all our major cities and explode them simultaneously, thus destroying a large part of our population, aggregated in densely populated metropolitan districts. Our possibilities of retaliation—even if retaliation should be considered adequate compensation for the loss of millions of lives and the destruction of our largest cities—will be greatly handicapped because we must rely on aerial transportation of the bombs and also because we may have to deal with an enemy whose industry and population are dispersed over a larger territory.

In fact, if the race for nuclear armaments is allowed to develop, the only apparent way in which our country can be protected from the paralyzing effects of a sudden attack is by dispersal of those industries which are essential for our war effort and dispersal of the populations of our major metropolitan cities. As long as nuclear bombs remain scarce (i.e., as long as uranium remains the only basic material for their fabrication), efficient dispersal of our industry and the scattering of our metropolitan populations will considerably decrease the temptation to attack us by nuclear weapons.

At present, it may be that atomic bombs can be detonated with an effect equal to that of twenty thousand tons of TNT.[1] One of these bombs could then destroy something like three square miles of an urban area. Atomic bombs containing a larger quantity of active material but still weighing less than one ton, which could destroy over ten square miles of a city, may be expected to be available within ten years. A nation able to assign ten tons of atomic explosives for a sneak attack on this country can then hope to achieve the destruction of all industry and most of the population in an area from five hundred square miles upward. If no choice of targets, with a total area of five hundred square miles of American territory, contains a large enough fraction of the nation's industry and population to make their destruction a crippling blow to the nation's war potential and its ability to defend itself, then the attack will not pay and may not be undertaken. At present, one could easily select in this country a hundred areas of five square miles each whose simultaneous destruction would be a staggering blow to the nation. Since the area of the United States is about three million square miles, it should be possible to scatter its industrial and human resources in

[1] This was an "educated guess" that proved correct at Alamogordo (New Mexico), Hiroshima, and Nagasaki.

such a way as to leave no five hundred square miles important enough to serve as a target for nuclear attack.

We are fully aware of the staggering difficulties involved in such a radical change in the social and economic structure of our nation. We felt, however, that the dilemma had to be stated, to show what kind of alternative methods of protection will have to be considered if no successful international agreement is reached. It must be pointed out that in this field we are in a less favorable position than the nations which are now more diffusely populated and whose industries are more scattered, and the nations whose governments have unlimited power over the movement of their populations and the location of industrial plants.

If no efficient international agreement is achieved, the race for nuclear armaments will be on in earnest not later than the morning after our first demonstration of the existence of nuclear weapons. After this, it might take other nations three or four years to overcome our present head start and eight or ten years to draw even with us if we continue to do intensive work in this field. This might be all the time we would have to bring about the relocation of our population and industry. Obviously, no time should be lost in inaugurating a study of this problem by experts.

PROSPECTS OF AGREEMENT

The consequences of nuclear warfare, and the type of measures which would have to be taken to protect a country from total destruction by nuclear bombing, must be as abhorrent to other nations as to the United States. England, France, and the smaller nations of the European continent, with their congeries of people and industries, would be in a particularly desperate situation in the face of such a threat. Russia and China are the only great nations at present which could survive a nuclear attack. However, even though these countries may value human life less than the peoples of western Europe and America, and even though Russia in particular has an immense space over which its vital industries could be dispersed and a government which can order this dispersion the day it is convinced that such a measure is necessary, there is no doubt that Russia, too, will shudder at the possibility of a sudden disintegration of Moscow and Leningrad, almost miraculously preserved in the present war, and of its new industrial cities in the Urals and Siberia. Therefore, only lack of mutual trust, and not lack of desire for agreement, can stand in the path of an efficient agreement for the prevention of nuclear warfare. The achievement of such an agreement will thus essentially depend on the integrity of intentions and readiness to sacrifice the necessary fraction of one's own sovereignty of all the parties to the agreement.

One possible way to introduce nuclear weapons to the world—which may particularly appeal to those who consider nuclear bombs primarily as a secret weapon developed to help win the present war—is to use them without warning on appropriately selected objects in Japan.

Although important tactical results undoubtedly can be achieved by a sudden introduction of nuclear weapons, we nevertheless think that the question of the use of the very first available atomic bombs in the Japanese war should be weighed very carefully, not only by military authorities, but by the highest political leadership of this country. Russia, and even allied countries that bear less mistrust of our ways and intentions, as well as neutral countries, may be deeply shocked by this step. It may be very difficult to persuade the world that a nation which was capable of secretly preparing and suddenly releasing a new weapon, as indiscriminate as the German rocket bombs and a thousand times more destructive, is to be trusted in its proclaimed desire of having such weapons abolished by international agreement. We have large accumulations of poison gas, but do not use them, and recent polls have shown that public opinion in this country would disapprove of such a use even if it would accelerate the winning of the Far Eastern war. It is true that some irrational element in mass psychology makes gas poisoning more revolting than blasting by explosives, even though gas warfare is in no way more "inhuman" than the war of bombs and bullets. Nevertheless, it is not at all certain that American public opinion, if it could be enlightened as to the effect of atomic explosives, would approve of our own country's being the first to introduce such an indiscriminate method of wholesale destruction of civilian life.

Thus, from the "optimistic" point of view—looking forward to an international agreement on the prevention of nuclear warfare—the military advantages and the saving of American lives achieved by the sudden use of atomic bombs against Japan may be outweighed by the ensuing loss of confidence and by a wave of horror and repulsion sweeping over the rest of the world and perhaps even dividing public opinion at home. From this point of view, a demonstration of the new weapon might best be made, before the eyes of representatives of all the United Nations, on the desert or a barren island. The best possible atmosphere for the achievement of an international agreement could be established if America could say to the world, "You see what sort of a weapon we had but did not use.

We are ready to renounce its use in the future if other nations join us in this renunciation and agree to the establishment of an efficient international control." After such a demonstration the weapon might perhaps be used against Japan if the sanction of the United Nations (and of public opinion at home) were obtained, perhaps after a preliminary ultimatum to Japan to surrender or at least to evacuate certain regions as an alternative to their total destruction. This may sound fantastic, but in nuclear weapons we have something entirely new in order of magnitude of destructive power, and if we want to capitalize fully on the advantage their possession gives us, we must use new and imaginative methods.

It must be stressed that if one takes the pessimistic point of view and discounts the possibility of an effective international control over nuclear weapons at the present time, then the advisability of an early use of nuclear bombs against Japan becomes even more doubtful, quite independently of any humanitarian considerations. If an international agreement is not concluded immediately after the first demonstration, this will mean a flying start toward an unlimited armaments race. If this race is inevitable, we have every reason to delay its beginning as long as possible in order to increase our head start still further.

The benefit to the nation, and the saving of American lives in the future, achieved by renouncing an early demonstration of nuclear bombs and letting the other nations come into the race only reluctantly, on the basis of guesswork and without definite knowledge that the "thing does work," may far outweigh the advantages to be gained by the immediate use of the first and comparatively inefficient bombs in the war against Japan. On the other hand, it may be argued that without an early demonstration it may prove difficult to obtain adequate support for further intensive development of nucleonics in this country and that thus the time gained by the postponement of an open armaments race will not be properly used. Furthermore, one may suggest that other nations are now, or will soon be, not entirely unaware of our present achievements and that consequently the postponement of a demonstration may serve no useful purpose as far as the avoidance of an armaments race is concerned and may only create additional mistrust, thus worsening rather than improving the chances of an ultimate accord on the international control of nuclear explosives.

Thus, if the prospects of an agreement are considered poor in the immediate future, the pros and cons of an early revelation of our possession of nuclear weapons to the world—not only by their actual use against Japan, but also by a prearranged demonstration —must be carefully weighed by the supreme political and military leadership of the country, and the decision should not be left to the considerations of military tactics alone.

It may be pointed out that scientists themselves have initiated the development of this "secret weapon," and it is therefore strange that they should be reluctant to try it out on the enemy as soon as it is available. The answer to this question was given above—the compelling reason for creating this weapon with such speed was our fear that Germany had the technical skill necessary to develop such a weapon and that the German government had no moral restraints regarding its use.

Another argument which could be quoted in favor of using atomic bombs as soon as they are available is that so much taxpayers' money has been invested in these projects that the Congress and the American public will demand a return for their money. The attitude of American public opinion, mentioned earlier, in the matter of the use of poison gas against Japan shows that one can expect the American public to understand that it is sometimes desirable to keep a weapon in readiness for use only in extreme emergency; and as soon as the potentialities of nuclear weapons are revealed to the American people, one can be sure that they will support all attempts to make the use of such weapons impossible.

Once this is achieved, the large installations and the accumulation of explosive material earmarked at present for potential military use will become available for important peacetime developments, including power production, large engineering undertakings, and mass production of radioactive materials. In this way, the money spent on wartime development of nucleonics may become a boon for the peacetime development of the national economy.

Methods of International Control

We shall now consider the question of how an effective international control of nuclear armaments can be achieved. This is a difficult problem, but we think it soluble. It requires study by statesmen and international lawyers, and we can offer only some preliminary suggestions for such a study.

Given mutual trust and willingness on all sides to give up a certain part of their sovereign rights by admitting international control of certain phases of the national economy, the control could be exercised (alternatively or simultaneously) on two different levels.

The first and perhaps simplest way is to ration the raw materials, primarily the uranium ores. Production of nuclear explosives begins with the processing of uranium in large isotope separation plants or huge production piles. The amounts of ore taken out of the ground at different locations could be controlled by resident agents of the international control board, and each nation could be allotted only an amount which would make large-scale separation of fissionable isotopes impossible.

Such a limitation would have the drawback of also making impossible the development of nuclear power for peacetime purposes. However, it need not prevent the production of radioactive elements on a scale sufficient to revolutionize the industrial, scientific, and technical uses of these materials and would thus not eliminate the main benefits which nucleonics promises to bring to mankind.

An agreement on a higher level, involving more mutual trust and understanding, would be to allow unlimited production but to keep exact bookkeeping on the fate of each pound of uranium mined. If in this way a check is kept on the conversion of uranium and thorium ore into pure fissionable materials, the question arises as to how to prevent accumulation of large quantities of such materials in the hands of one or several nations. Accumulation of this kind could be rapidly converted into atomic bombs if a nation should break away from international control. It has been suggested that a compulsory denaturation of pure fissionable isotopes may be agreed upon—by diluting them, after production, with suitable isotopes to make them useless for military purposes, while retaining their usefulness for power engines.

One thing is clear: any international agreement on prevention of nuclear armaments must be backed by actual and efficient controls. No paper agreements can be sufficient, since neither this nor any other nation can stake its whole existence on trust in other nations' signatures. Every attempt to impede the international control agencies would have to be considered equivalent to denunciation of the agreement.

It hardly needs stressing that as scientists, we believe that any systems of control envisaged should leave as much freedom for the peacetime development of nucleonics as is consistent with the safety of the world.

SUMMARY

The development of nuclear power not only constitutes an important addition to the technological and military power of the United

States but also creates grave political and economic problems for the future of this country.

Nuclear bombs cannot possibly remain a "secret weapon" at the exclusive disposal of this country for more than a few years. The scientific facts on which their construction is based are well known to scientists of other countries. Unless an effective international control of nuclear explosives is instituted, a race for nuclear armaments is certain to ensue following the first revelation of our possession of nuclear weapons to the world. Within ten years other countries may have nuclear bombs, each of which, weighing less than a ton, could destroy an urban area of more than ten square miles. In the war to which such an armaments race is likely to lead, the United States, with its agglomeration of population and industry in comparatively few metropolitan districts, will be at a disadvantage compared with nations whose population and industry are scattered over large areas.

We believe that these considerations make the use of nuclear bombs for an early unannounced attack against Japan inadvisable. If the United States were to be the first to release this new means of indiscriminate destruction upon mankind, they would sacrifice public support throughout the world, precipitate the race for armaments, and prejudice the possibility of reaching an international agreement on the future control of such weapons.

Much more favorable conditions for the eventual achievement of such an agreement could be created if nuclear bombs were first revealed to the world by a demonstration in an appropriately selected uninhabited area.

In case chances for the establishment of an effective international control of nuclear weapons should have to be considered slight at the present time, then not only the use of these weapons against Japan, but even their early demonstration, may be contrary to the interests of this country. A postponement of such a demonstration will have in this case the advantage of delaying the beginning of the nuclear armaments race as long as possible.

If the government should decide in favor of an early demonstration of nuclear weapons, it would then have the possibility of taking into account the public opinion of this country and of the other nations before deciding whether these weapons should be used against Japan. In this way, other nations may assume a share of responsibility for such a fateful decision.

PART II · TAKING STOCK

Introduction to Part II

The second part of this book differs from the first in emphasis; articles reprinted here attempt to sum up and interpret what has already happened rather than to look into the future. Several of them were written on anniversaries, New Year's days, or other occasions which called for retrospection. The period covered by these articles includes, in political history, the Soviet seizure of Czechoslovakia in 1948, which put an end to illusions about postwar relations between wartime allies; the Communist victory in China in 1949; the Korean War in 1950; and the attack on Suez and the revolution in Hungary in 1956. It covers seven years of Truman and eight years of Eisenhower administration, the death of Stalin and the rise of Khrushchev. In military technology, this period has brought about two major developments—the replacement of the fission bombs, which wiped out Hiroshima and Nagasaki, by the much more destructive fusion bombs, and the development of ballistic missiles, which made every point on the globe open to nuclear attack from every other point.

In 1956, in reviewing the developments since 1945, I called these years "Ten Years that Changed the World." The scope and speed of changes have not slowed down in the years that followed. They have witnessed, in politics, the explosive emancipation of the colonial world, and in technology, the first space flights. Only some aspects of these breath-taking events are reflected in the articles that follow. Similarly to the rest of the book, they are dominated by the conviction that revolutionary changes in science and technology are the unique and decisive feature of the transformations which mankind is undergoing in our time, and their emphasis is on this aspect of contemporary history.

Undoubtedly, the rise of communism and its challenge to the economic system developed in Western Europe and in the United States is a major historical event; the awakening of the peoples of Africa and Asia, the collapse of the great colonial empires, and the end of the belief in permanent superiority of the white race are phenomena of even greater significance. However, only the coincidence between the social and national revolutions of our time, on the one hand, and the explosive advancement of science and technology, on the other, gives these revolutions a critical importance quite different from that of revolutionary changes in the past. Without the scientific revolution, the contest for power between the Soviet Union and the United States would have been settled, sooner or later, by the traditional mechanism of war; but the technological revolution makes war irrational and brings the conflict to a dead end. Many oppressed or colonial nations have thrown off alien rule in the past; new ones would have done it in our time, irrespective of the technological revolution. Most of them would have merely followed in the steps of Kemal's Turkey or Toussaint L'Ouverture's Haiti, adding a few new nations to the traditional list. It is the scientific revolution that converts the traditional phenomenon of the rise of new and the decline of old nations into a world-wide "revolution of rising expectations."

The scientific revolution has transformed the familiar power struggles between military camps into a contest for technological supremacy. It is now extending this contest from means of inflicting destruction in war to means of promoting the industrialization of nations in peace. It has given the competition between the two power blocks the abstruse form of a "race to the moon."

The new technological abilities make old Adam a new man; they call for transformation of his traditional competing societies into a single world-wide community as the only alternative to mutual destruction.

9

Two Years after Hiroshima

Introduction

In September, 1947, two years had passed since the atom bomb blast over Hiroshima. To all appearances, American monopoly of nuclear weapons stood unchallenged; but behind walls of secrecy, better guarded then the secrecy of the Manhattan project in wartime, Professor Igor Kurchatov and his crew were working feverishly to break this monopoly and were apparently nearing success. In the United Nations Atomic Energy Committee, negotiations on international control of atomic energy, to which the East and the West had agreed in principle, dragged on. The Soviet Union, bent on gaining time for its own weapon development, retreated slowly from its initial proposal to "ban the bomb and talk about controls later." It began to acknowledge the need for reliable controls of the ban, to be agreed upon before it would take effect; but these concessions were much too vague for a meeting of minds. The American leadership, on the other hand, in the mistaken belief in a lasting United States monopoly of nuclear weapons, refused to specify the time when American nuclear weapons would be eliminated. The clock of the nuclear arms race was ticking along, but few people were aware of it and still fewer alarmed by it.

The developments of the last two years have been contradictory, in the field of atomic energy as in the wider realm of human affairs. The belief that the sovereign national state is obsolete, and a community of mankind under universal law is possible and feasible *now,* gathers thousands of enthusiastic proponents and millions of passive supporters. Yet, at the same time, militarism rises in traditionally civilian America, and nationalism in the Soviet Union, which was originally dedicated to intransigent internationalism. Generals and admirals admit that national security is an obsolete notion, but military budgets swell as never before.

The atomic bomb has much to do with these contradictory trends. Opinion polls show that a substantial majority of Americans are ready for extreme sacrifices of national sovereignty, if these could buy freedom from fear of an atomic war. The necessity for international control of atomic energy is universally recognized. The American Legion favors it now and calls for strengthening the United

From *Bulletin of the Atomic Scientists,* September, 1947.

Nations; General Leslie R. Groves (previously head of the Manhattan project) supports it and asks for abolition of war. Perhaps, some conversions to international rule over atomic energy and a veto-free United Nations have been made easier by the Soviet Union's opposition to both. But fundamentally, these conversions reflect the recognition that nothing short of a world-wide authority, invested with real power, can bar the road of the Western world to self-destruction. At the same time, the feeling of overwhelming power which the bomb gives the United States contributes largely to the callous nonchalance with which a new war is often discussed. Apprehension of the time when America will not be the sole possessor of the Absolute Weapon breeds thoughts of a preventive war. The fear of losing atomic secrets endangers the fabric of American civil liberties.

The American policy since Hiroshima has exhibited the same antinomy. The American government has sponsored a bold and enlightened plan for the solution of the atomic energy problem. For over a year now the American delegation has persevered in trying to obtain Soviet consent to this plan. We believe that if this plan were to be adopted by the United Nations, the Senate would ratify it, because enough senators would be convinced of its necessity and because the pressure of public opinion, which could be marshaled in its support, would be much stronger than that which defeated the May-Johnson bill and forced the confirmation of Mr. Lilienthal as chairman of the Atomic Energy Commission.

And yet, we have recently allowed our unassailable position to be compromised, and the weak position of Russia strengthened, by our vagueness on the relatively minor question of the disposal of the existing atomic bombs. We want a world free of the threat of atomic weapons; but we also want to keep our bombs as long as possible. If international control is established, we might still want a few bombs to be left at the disposal of the United Nations, in the hope that if the worst comes to the worst these bombs will be "on our side."

In the general field of world politics, the American policy also has a double set of objectives. We work both for the strengthening of the United Nations and for the acquisition of outlying bases, for universal disarmament and for secret development of new methods of mass destruction, for rehabilitation of war-devastated Europe and Asia and for a distribution of our assistance that would strengthen our allies and weaken our potential enemies. In brief, we try to abolish war and to be in the position to win it.

The dilemma is a universal one. The representatives of ten nations in the United Nations Atomic Energy Commission have embraced the Baruch-Lilienthal plan without protesting the infringement of their nation's sovereignties. Among them was Mr. van Kleffens, the same Dutch representative who recently argued that the United Nations has no business to interfere with the sovereignty of the Netherlands in Indonesia. Holland is for the universal rule of law—but also for the right to engage in a private war to re-establish Dutch colonial power.

Is not the Soviet attitude, at least, free of this antinomy? Has not the Russian rejection of the Baruch plan "in whole or in part" clearly demonstrated that the Soviet Union wants no part of a supranational authority?

Since the negotiations began fourteen months ago, the Russians have conceded, step by step, that international control of atomic energy is necessary and should include both atomic armaments and atomic power for peaceful purposes, that it should be exercised by an international personnel having unrestricted access to all mines, plants, and laboratories engaged in atomic activities, that the control agency must sponsor international research, and that the atomic energy developments in each country should be subject to a quota system.

The Russian-proposed system is still far from satisfactory. Two crucial questions remain to be answered: First, how does the Soviet Union propose to satisfy other countries that no illegal mining and processing occurs on its territory? Second, can efficient (but not "unlimited") inspection be organized without international management of major atomic activities, as envisaged in the Lilienthal plan? We do not know how Gromyko will answer these questions, or whether he will answer them at all. It seems likely that the UN Commission is headed not for a compromise but for another majority report. Nevertheless, it is fair to say that during the last year even the U.S.S.R. has been moving toward the recognition of the need for supranational authority in the field of atomic energy, while at the same time moving toward isolationism and unilateral policies in its general relation with the world.

The desire of the peoples for peace and the dread which even the most ruthless rulers must feel in envisaging the probable consequences of a new war push all nations toward a community of interests and efforts. Modern technology makes mockery of any attempt to perpetuate the division of the world, whether into a dozen or in two fragments. No mind, however blinded by indoctrination, can fail to recognize these facts. And yet, there seems to be no way of stopping the inertia of a world system in which wars are inevitably recurrent events.

Because of this inertia, the labors of the United Nations Atomic Energy Commission begin to appear vain, almost ghostlike. Even if this commission would unanimously recommend to the Security Council an effective system of control, how could its provisions be put into operation under the present world conditions? The solution of the problem of atomic energy has become contingent on a change in the general political and economic relations between the nations of the world. Two years ago many hoped that the blinding flash of Hiroshima would frighten mankind onto the paths of peace; one year ago many of us hoped that mankind could be enticed to enter the edifice of peace through a little back door inscribed ADA.[1] Today this hope is all but gone.

It is easy to say: The situation is not of American making, and we are powerless to change it. Our plans for internationalization of atomic energy as well as our appeal for participation in a common effort to reconstruct Europe have fallen on deaf ears. How can we stop preparing for war and strengthening our alliances when a first-rate military power refuses to participate effectively in the world peace organization and insists on surrounding itself with satellites?

Each of our proposals may be fair and justifiable by circumstance —the Baruch-Lilienthal plan, the Truman doctrine, the Marshall plan. But our basic political philosophy remains that of all national states—that the security and prosperity of a nation are the supreme good, and their attainment, the only legitimate aim of a national government. This philosophy debases the most enlightened policies. When the security and prosperity of other peoples are promoted, a reason—or rather an excuse—is proffered that it best serves our own national interests. The atomic scientists had assumed that the only way to sell international control was to prove that it is needed for American security. The Marshall plan has been presented as serving the ultimate aim of assuring American economic stability and military strength.

Since even the most humanitarian policies are supposed to derive their validity from their subordination to the paramount criterion of selfish national advantage, they are inevitably reduced to the role of new, untried, and therefore unreliable weapons in an arsenal full of reliable stand-bys: big armies, navies, and air forces; fortresses and naval bases; loans with political strings attached. And equally inevitably, they are acknowledged by the other nations of the world for what we ourselves proclaim them to be—measures intended primarily to serve the interests of the United States of America.

[1] Atomic Development Authority, proposed by the United States as a UN-operated agency monopolizing all large-scale production and handling of fissionable materials.

It would not be enough if we were to start pretending that whatever we do in the world is for the benefit of mankind as a whole; we should really make this benefit our paramount objective. The world is tired of hearing every nation assert that its own strength is the best guarantee of world peace. We must convince ourselves, before we can convince others. Not until we ourselves believe that equal concern with the well-being and peace of all mankind is the only possible moral basis for world leadership, will we stand a chance of acquiring this leadership and breaking the vicious circle of power politics.

10

Europe in July, 1954

Europe Returns to Normalcy

To formulate impressions after a brief visit to a foreign country is a presumption. It is even more dangerous to generalize such fleeting and scattered impressions to cover a multinational area such as Western Europe. Nevertheless, here, for what they may be worth, are a few observations gathered while visiting Europe this summer, listening to people in different countries and walks of life, reading newspapers and journals, and trying to acquire a feel for the climate of public life.

For the first time since the war, there exists in Europe a rather widespread feeling of relative security and prosperity, not only in rich and smug Switzerland or booming West Germany, but also in England, emerging from the drabness of self-imposed austerity, and in France, believing that she has found the first stable and popular government after nine crisis-ridden years. The national economies of most countries of Western Europe have acquired a certain degree of stability and independence from dollar help. Reconstruction—new housing, new factories—is noticeable everywhere, even in long-stagnant France, not to speak of Germany, which is one great construction site. The tempo of scientific research is accelerating, as war-destroyed or depopulated laboratories are rebuilt and staffed with a new generation of scientists. The fear of war has receded. Since the centers of world tension have moved to the Middle East and southeast Asia, the European nations feel relief from the strain under which they lived when alarm signals flew over Berlin and Belgrade. People don't want to hear about the threat of a new war, atomic or otherwise. They have had more than enough of crises and emergencies and want to live their lives as best they can. They cannot deny that the present economic and political stability is precarious, but it is the best they have had for years, and they are set to enjoy it.

In this continent of tired nations, who are just finding again the joy of living, America appears as a doctrinaire kill-joy. Not many people in Europe believe (although some do) that America is deliberately seeking a new war; but it is enough that America is trying

From *Bulletin of the Atomic Scientists,* October, 1954.

to maintain in Europe a feeling of emergency, urging a sustained and increased defense effort, to make her unpopular. In the striving for relaxation, traditional values are upset. The settlement in Indochina, however detrimental to French political and military prestige, was widely received in France not merely as a relief (which was understandable enough) but as a vicarious triumph—a victory over a conspiracy to keep France entangled in a faraway disastrous war. The French reaction to the Allied pressure for the ratification of the European Defense Community can be traced to the same attitudes. To participate in the EDC would have meant a grave commitment; it could have involved the country in quarrels outside its narrow national interests, such as the almost inevitable conflict over Germany's eastern lands. What a tempting occasion to assert France's independence and at the same time to indulge the urge for normalcy! What satisfaction to flaunt the admonitions of powerful (and sometimes overbearing) friends, on whose help one is not as dependent now as one has been in past years, and in doing so, to escape acknowledging an emergency necessitating painful decisions and strenuous efforts! The French people's reaction to the EDC was, in essence, *fichez-nous la paix*—leave us alone!

Of course, the Western European isolationist and neutralist mood is even less justifiable in fact and likely to prove even shorter-lived than the complacent moods of America and England in the aftermath of the First World War. But, for the moment, the newly found freedom of choosing is intoxicating, and the choice of inaction is sweet. National emergencies, patriotic duties, appeals for greatness and sacrifice, have played havoc with the lives of two European generations. Europe has had enough of them; and in 1954, her inclination (carefully abetted by the new Soviet policy) is toward stabilization, normalcy, and refusal to be alarmed. In a sense, this is a healthy sign of recovered self-confidence, conducive to economic progress and creative work in all fields; but at the same time, what an ominous reminder it is of the fool's paradise in which England lived under Baldwin and Chamberlain and America under the Neutrality Act.

Of course, many responsible political leaders, and simple people, in France, England, and elsewhere in western Europe are clearly aware of all the dangerous implications of the world situation. The motives of those (non-Communist) French deputies who voted down the EDC were much more complex, individual, and rational, than mere indulgence in the desire for normalcy; but without the strong popular swell of the mood for relaxation, the outcome of the controversy might well have been different.

THE LAVA SETS

Immediately after a great war, everything is in flux. The old forms of national organizations are discredited; peoples are in the mood to forsake them and try something new, to avoid the repetition of past ordeals. They are willing to listen to new prophets. After the First World War, President Wilson appeared as such a prophet; in 1918, he had the ear and trust of European nations to a degree of which he himself was hardly aware. (At Versailles, he talked to politicians and not to peoples.) But instead of American leadership in the political reconstruction of the world, what came was a resurrection of American isolationism.

For a second time, in 1945, the nations of Europe were ready, with even more trust than in 1918, to follow American leadership into a new world. Even the Soviet rulers would have found it difficult, at that time, to stay away and denounce as a capitalist plot a truly imaginative American program for effective elimination of war. But no bugle call ushering the inception of a new world was heard from Washington. Instead, in Yalta and Potsdam, America sought to lay a diplomatic foundation for a political withdrawal from Europe, militarily already in full swing. Treaties were signed guaranteeing mutual respect for the respective spheres of influence, and assurances of "independence" were given to nations entrapped in the Soviet mesh.

The American withdrawal from Europe that followed these spurious settlements lasted only two or three years, but the most propitious moment for decisive action for peace was lost in this interlude. The military inferiority and economic disorganization, which had threatened the Soviet system with collapse immediately after the war, and the gratitude of the Russian masses for America's help in the war, had offered, in 1945, at least a fighting chance to entice—or force—the Soviet Union into partnership in a really effective world organization. This was never attempted. Instead, the old league of sovereign nations was revived under a new name. American participation in such a league would have made all the difference in 1919, but was "too little and too late" after the birth of the atomic bomb and the emergence of the Soviet Union as a world power. No attempt was made to shape, while the iron was still hot, solutions of the basic problems of political security and economic stability which beset the world. Instead, controversial questions, such as Germany's eastern boundaries and her ultimate international status, were postponed indefinitely on the plausible—but almost always false—premise that such problems would be easier to solve when things had settled and tempers cooled down. Similarly, the endowment of the United Nations with a permanent military force to enforce peace was left to future negotiations. The immense asset of a fully mobilized American military power and of an American industrial economy working in high gear while that of the rest of the world was largely crippled—the asset which had been so effectively used to win the war—was not brought to bear, in the decisive moment, to ensure lasting peace.

American re-emergence in European affairs came in time to permit economic and political rehabilitation of Western Europe but too late to prevent the rebirth and solidification of an isolationist Soviet empire. By 1948, the atomic bomb was no more an American monopoly; the worst economic disaster in the Soviet Union had been averted, and the Soviet leaders had reacquired enough freedom of action to reject aid under the Marshall Plan, and to resist the United Nations plan of sharing atomic energy under international management, denouncing it as a capitalist conspiracy to subvert the Soviet economy.

The moral and economic instability of Western Europe has been slower to disappear than the Soviet weakness. Under the restless urging of a dictatorship, the Soviet Union returned to what can be grimly called "Soviet normalcy" several years before Western Europe began to regain her strength. The rough aggressiveness of the newly invigorated Soviet policy (culminating in the subversion and subjection of Czechoslovakia) awakened Western Europe to its own continued weakness and to the inadequacy of its traditional multinational setup. This gave America a second chance in three years to lead in the reorganization of the world for peace, although this time, only that part of the world which had remained outside the sway of the Kremlin could be encompassed. The first steps to meet this new challenge—the Marshall Plan, and the Point Four program of aid to underdeveloped countries—were bold and effective. The first brought Western Europe out of a critical economic and political depression; the second gave new hope to large areas of the world. A good beginning was made, but no follow-up came. The vision of American statesmen did not go beyond emergency help to tide one or another nation—old friend or recent enemy—over a difficult time. The Marshall Plan had been sold to the American people, not as a first step toward a permanent economic and political integration of the West, but as a terminal action, which—it was promised—would turn the clock back and make European nations able to again in the past. With this unimaginative aim-setting, it was inevitable take care of their economic and political affairs as they used to do that the more successful the Marshall Plan became, the more effectively it put an end to American political influence in Europe. A permanent active partner retains his influence in an enterprise; a rescuer in hard times is bound to lose his when the emergency passes, whatever the gratitude owed to him.

The molten lava, left after the volcanic eruption of the Second World War, which America could have molded almost at will in 1945, is now solidifying into familiar, or new, rigid shapes of national states, each preoccupied with its own internal problems and distrustful of most, if not all, of the others. Men have staked out their lots to till the new fields; a generation is growing up which takes the new division of the world for granted (they learn it at school!) and wishes to be left alone and undisturbed to build their economic and intellectual homes on the apparently solid ground, unmindful of the danger of a new eruption. National antagonisms and hatreds, which seemed for awhile to be dissolving in the crucible of common suffering, emerge again as the suffering subsides.

Germany is slowest to get into this mood, because its division into two zones is too grotesque—and for Germans, too revolting—to rapidly become a familiar and therefore legitimate feature of the European map. It is difficult to imagine how even a second or a third generation of Germans could grow into the acceptance of this division as a permanent form of national existence. This is why Germany, more than any other country in Europe, sympathizes with the American concept of permanent emergency in relations with the Soviet Union, a temporary community of feeling which some Americans mistake for a lasting community of ideas. However, barring a political upheaval and collapse of the Soviet empire, one does not see how even the division of Germany can now be ended without a new war. One must therefore expect that in this area, too, the presently existing structure, however absurd it appears now, will survive and stiffen with time.

AMERICAN LEADERSHIP WANES

As its once fluid substrate becomes more viscous, and more difficult to mold, foreign policy calls for increasing patience and skill. Inspiration and rapid decision could have reshaped the world in flux; but now, only patient diplomatic craftsmanship can accomplish even the slightest desired change (such as the settlement of the Trieste issue).

Although no grand design in world policy can now be realized in short order, it is of the greatest importance that such a design should exist and be known. If in America, as well as in all other countries, day-to-day political decisions and actions are guided by no exalted ultimate aim but only by narrowly interpreted, momentary national interests, this will inevitably lead (as it has always led in the past) to a sequence of more or less acute international tensions, one of which will ultimately erupt in war.

These criteria—the existence and adequacy of a "grand design" for peace and the wisdom and skill of day-to-day pronouncements and actions—must be used in judging American foreign policy (about which every American visitor gets an earful of criticism in Europe). In justice, it cannot now be said—as it has been so often asserted in the past—that America "has no foreign policy," meaning that her actions on the world scene are based on no general concept of the world situation and no over-all plan to deal with it. In this respect, American foreign policy under John Foster Dulles can claim a superiority over the policy practiced or recommended by some of its European critics—a policy of "muddling through" or "playing by ear," without a logically and historically defensible long-range perspective.

After the chance of bringing the Soviet Union, by persuasion or threat, into an effective supranational peace organization had been lost, without having been properly explored at the end of the war, the division of the world into two camps—two competing empires which are also two protagonists of conflicting political ideologies—has become a fundamental fact on which all long-range political planning has to be based. Recognizing this situation, and making it the basis of its policy, America is showing itself more realistic than those of its critics who would like to believe (or at least, to get others to believe) that a little good will, and some concessions here and there, could convert a gigantic, historical contest into a friendly "coexistence."

While present American policy cannot be fairly accused of the absence of a grand design (or a fundamental error in the evaluation of the world situation on which this design is based), the accusation that this design lacks imagination and is without sufficiently strong and enduring appeal to the world, often heard from thoughtful critics of American foreign policy in Europe, is not so easy to contradict. This design consists in surrounding the Soviet Union by a military alliance, or alliances, of sovereign nations armed as heavily as possible and maintaining these alliances as a deterrent to Soviet aggression for an indefinite time. The imaginative idealism on a world-wide scale professed (and practiced at least to some extent) by the preceding administrations has been consigned to the junk heap as a failure, since this policy did not cause the Soviet empire to collapse or the non-Communist nations to line up behind America as eternally grateful retainers. Idealism and imagination have gone out of fashion in Washington. The now predominant "political realism" of the State Department and Congress is set to terminate the foreign assistance programs in Europe and the Point Four programs in Asia and to restrict aid to military items, supplied only to nations willing to "play ball" militarily with America. Critics of this policy say that to combat the danger of a new, and unimaginably destructive, war it is important not only to bring into a military alliance the largest possible number of—willing or unwilling—European and Asiatic nations but, above all, to gain popular support for an American "grand design" for peace. Military alliances are formed in the face of threatening aggression; they are apt to disintegrate when this danger becomes, or appears to become, less acute, as it seems now to many Europeans. To anchor the adherence of nations to American leadership in more solid ground—the spontaneous allegiance of their peoples—the American program cannot be purely a negative one of defense against Soviet aggression. Rather, it must be an inspiring positive program, a blueprint of a world from which war is permanently eliminated. If there is to be a successful crusade, it must be one in search of the Holy Grail and not for the destruction of Carthage!

Achieving the ultimate aim of a world safe from war is outside our grasp at the present time; but dedicating ourselves to it is possible at any time. It was America as much as the Soviet Union that shied away in 1945 from sacrificing too much of its (largely imaginary) sovereignty. It is still open to us to abandon this fateful pusillanimity. To make the interest of all mankind rather than selfish national advantage the guiding star of American policy, to proclaim and accept our duty to become loyal members of the world community, would do more to keep the non-Communist world from falling apart than attempts to unite, in a military alliance, sovereign states mindful only of their own advantage.

Many will deride as utterly unrealistic the proposition that the creation of a world system of law can be made the aim of American world policy, or that the betterment of the conditions of life of all mankind can be made the aim of American economic policy, or that such aim-setting, if possible, would significantly help to deter the threatening aggression of Communist totalitarianism. These skeptics will assert that faced with the rapidly growing threat of atomic and hydrogen bombs, submarines, jet planes, and guided missiles, wielded by a group of ruthless fanatics bent on securing the world for their dogma (and incidentally, for their personal domination), the only effective policy is to weld the non-Communist world into a powerful military alliance. These "realists" discount the binding power of common ideals compared with that of standardized rifles.

Even to those who consider the present American world political design as the only one possible it must be obvious from the European point of view that, of late, the means used by American diplomacy to forward this design have been extremely clumsy and signally unsuccessful. Within two or three years, most of the good will toward America that had accumulated in various parts of the world under the previous administrations has been squandered. To keep the allegiance of European nations now, after they have regained their political self-reliance, calls for increased caution and tolerance. Instead, the Europeans say, Washington's accent has become more harsh and impatient. While the lessened economic dependence of European countries on America calls for strengthening the bonds of common ideals, American leadership has permitted the traditional admiration for America as the land of all parties and opinions to be practically wiped out by permitting interference with the individual freedoms and the right of dissent in America far in excess—people in Europe believe—of all rational requirements of national security.

If there is one impression an American brings back from a trip to Europe in 1954—an impression so consistent that it can safely be generalized to all European countries and all political groups (with the possible exception of the remnants of fascism and Nazism)—it is bewilderment over happenings such as the McCarthy-Army controversy, the Oppenheimer affair, and the denial of visas to famous, and often politically neutral or even conservative, scientists. The recent prohibition of the Communist party, and the fact that this measure has been put over by those whom Europe considered the liberal element in American political life, undoubtedly has made this bewilderment still deeper. Of course, sensational reporting and Communist propaganda immensely exaggerate the "terror" under which Americans are supposed to live. One of my friends seriously suggested that my failure to answer his letters might have been due to a fear of corresponding with foreigners; others have expressed astonishment that the *Bulletin of the Atomic Scientists* is permitted to criticize official visa and security policies or oppose the rejection of Madame Joliot-Curie by the American Chemical Society! But even without such exaggerations, enough strange things have actually happened in America to make the Communists (and the non-Communist intellectual detractors of America) gloat and the staunchest friends of America in Europe wring their hands. The Communist propaganda about the suppression of free thought and political opposition in America finds much credence in Europe—because there are enough true facts to make it appear plausible. When, however, the same propaganda attempts to convince Europeans that American workers live in abject poverty and do not have enough to eat, nobody falls for it.

In the eyes of most Europeans, Americans have lost their sense of proportion in attempting to eradicate a largely non-existent danger of internal subversion. The greater distance of America from the Soviet empire, and the small number of Communists in our midst, make the threat of the world Communist empire appear particularly ominous and mysterious to Americans. Those who have lived with the Soviet power on their frontiers for many years have acquired a certain philosophical—and sometimes even skeptical—attitude toward its threat. Those who daily encounter Communists among their friends and colleagues, or simply in the street throng, cannot help seeing in them ordinary men with extreme and obstinate political ideas but otherwise with the same weaknesses (and often the same human virtues) as everybody else rather than sinister conspirators, as they appear to a majority of Americans.

It is as if the glaring light thrown on the world by rising Soviet power, has converted the peaceful American scene into an eerie landscape of black and white, in which the fine outlines and delicate shades are lost to the eye. Small, unimportant things become exaggerated out of all proportion, while significant things are lost in the shadow. How to explain otherwise the opinions in the Oppenheimer case of the majority of the Gray Board and of the AEC—liberal and fair-minded people all—in whose eyes an occasional visit in Paris with an old colleague who was once known as a Communist, and a twelve-year-old (and long ago admitted) attempt to shield this friend from being implicated in an espionage investigation, cast so wide and black a shadow on Oppenheimer's reputation that it conceals a lifetime of brilliant achievement and a service of immeasurable importance to the nation in time of need! American visitors to Europe have a hard time explaining this kind of thing to people in Europe, who see that their own countries are not brought anywhere near collapse by extending to Communist groups or individuals—often much more numerous than in America—the traditional protection and tolerance minority groups enjoy in democratic countries.

American diplomatic agents abroad have warned the State Department that the activities of Senator McCarthy and his fellow congressional investigators have done tremendous damage to American standing in the world. Every visitor to Europe can fully endorse these reports; but it must be added that the security policies of the American administration, in particular, the often senseless visa decisions, are causing as much dismay and criticism in Europe as the antics of McCarthy and his friends.

Some American policy statements in the field of military policy have created an even greater alarm and resentment in Europe. The speeches of Vice-President Richard M. Nixon, Senator William F. Knowland, and Admiral Arthur Radford, and some of President Eisenhower's own pronouncements, have succeeded in making many in Europe believe that, at least at the height of the Indochina crisis, the American leaders were seriously considering starting a big new war in Asia and forcing France to supply the manpower for it. The best (and truthful) answer an American visitor is able to give to these accusations is that, in his conviction, at no time was there a serious intention in Washington to go to war over Indochina and that all contrary statements were parts of a bluff which—the State Department must have believed—might have been successful if Britain had not refused to associate herself with it. This answer, even if accepted as true, is not of a nature to strengthen European belief in American political wisdom. Nothing America has done in recent times has so deeply injured the willingness of Europeans to accept her leadership as did this sabre-rattling bluff.

The clumsiness of American foreign policy in the last few years, the discredit cast on American democracy by the flourishing of what has become known all over the world as McCarthyism, and the stabilization of the internal political and economic systems of most European countries have combined to destroy the willingness of these nations to accept American leadership. In the present psychological climate, even the most enlightened American initiative in the reorganization of the world for peace is certain to run into great skepticism in Europe. To put it in the simplest, if the most extreme, form: hardly any nation in western Europe would be willing now to enter a federal union with America, whereas popular support for such a union would have been overwhelming in each of these nations ten years ago.

President Eisenhower, sensing this waning respect for American leadership in Europe, has admonished Americans to talk less of "leadership" and more of "partnership." It is, however, not the name but the spirit that matters. Leadership in the non-Communist world is imposed on America by the forces of history and geography; this is fully realized in Europe. But the aim toward which America shows the lead, and the wisdom, self-restraint, and skill she uses in exercising this leadership, will decide whether it will succeed or fail.

11

Ten Years That Changed the World

Man has mounted science, and is now run away with. I firmly believe that before many centuries more, science will be the master of man. The engines he will have invented will be beyond his strength to control. Some day science may have the existence of mankind in its power, and the human race commit suicide by blowing up the world.

HENRY ADAMS, 1862

THE CLOCK STARTS TICKING

With this number, the *Bulletin of the Atomic Scientists* enters its second decade, its first issue having appeared on December 15, 1945. It was founded by a group of scientists whose participation in the development of the atomic bomb convinced them that, with this discovery, a radical change had come in the role of science in public affairs. They believed that mankind was entering unawares into a new age fraught with unprecedented dangers of destruction. In the spring of 1945, this conviction led some scientists to an attempt—perhaps the first one in history—to interfere, *as scientists,* with the political and military decisions of the nation. Leo Szilard's memorandum to President Roosevelt (March, 1945) and the report to the Secretary of War by the so-called Franck Committee on June 11, 1945, counseling, for reasons of long-range policy, against the use of atomic bombs in Japan, were the first manifestations of this new concern of scientists with public policy.[1]

Their arguments made hardly a ripple in the flow of events that led to the holocausts of Hiroshima and Nagasaki.[2] The government was unprepared to see in the atom bomb anything but a bigger blockbuster, a welcome means for rapid termination of the siege of Japan. With few exceptions, public opinion rejoiced over Hiroshima

From *Bulletin of the Atomic Scientists,* January, 1956.

[1] Szilard's memorandum, "Atomic Bombs and the Postwar Position of the United States in the World—1945," was reprinted in *Bulletin of the Atomic Scientists,* III (December, 1947), 351–53; the "Franck Report" is reprinted as an addendum to Part I of this book.

[2] Henry L. Stimson, "The Decision To Use the Atomic Bomb," *Bulletin of the Atomic Scientists,* III (February, 1947), 37–44, 66–67.

and Nagasaki as demonstrations of American technical ingenuity and military ascendency. The certitude that other nations would soon acquire similar weapons and the vision of American cities crumbling in future "atomic Pearl Harbors," of which scientists spoke with deep apprehension, were discounted by the public as obsessions of a "little band of frightened men" whose consciences were troubled by their participation in the development of such an enormously destructive weapon.

The "atomic scientists" saw ahead a long task of public education to make the peoples of the world understand the challenge with which the atomic age confronted them: to establish permanent peace or face destruction of their civilization in an atomic war. The *Bulletin* was to be the voice by which these scientists could speak to the public as well as a forum in which they could discuss matters of public policy for their own enlightenment.

Senator Brien McMahon was one of the few statesmen who understood from the beginning the paramount importance of atomic energy for the future of mankind. The motto on the *Bulletin*'s masthead, "It is our solemn obligation . . . to lift our eyes above the lesser problems and to discuss and act upon . . . the supreme problem before our country and the world," is from a speech which he had prepared for delivery a few days before his untimely death. For most other national leaders—as for the public in general—the threat of the atom appeared remote beside the many urgent problems posed by day-to-day national and international developments.

COULD THE CLOCK HAVE BEEN STOPPED?

To prevent the atomic clock from striking the hour of doom, the idea of international control of atomic energy was conceived early by scientists. There never was—there never could have been—much hope that if large-scale production of nuclear materials was permitted to get under way in several nations enjoying essentially unrestricted sovereignty nuclear weapons could be effectively eliminated from their arsenals. In 1945, however, massive non-military atomic fuel production had not yet started; and it was not unreasonable to hope that rapid establishment of an international control system could forestall the diversion of atomic materials to the making of weapons and thus prevent a race in atomic armaments—or, at least, delay its beginning long enough for all nations to realize what changes in national sovereignties would be necessary to avoid this race.

From the point of view of gaining time for reflection, perhaps the most logical suggestion was that of a moratorium on all large-scale nuclear technology, for industrial power as well as for military purposes.[3] Such an agreement—much simpler to control than a prohibition of military applications of an otherwise freely growing atomic industry—would have stopped the atomic clock dead until the United Nations decided that the political setup of the world had become safe enough to give industrial atomic power the go-ahead signal.

A different approach was chosen in the Acheson-Lilienthal report,[4] which became the basis of American proposals to the United Nations. It envisaged a full-speed development of nuclear technology under the aegis of the UN. To make this development safe, the ownership of dangerous nuclear materials and the operation of all installations capable of producing them were to be vested in the international agency.

The "positive" vision of international co-operation in atomic technology kindled the imagination of scientists all over the world; the "negative" idea of delaying progress in the name of safety appeared unattractive beside it. Since American aversion to "stopping the wheels of progress" was more than matched by the adulation of technical progress in the Soviet Union, the "stop and think" proposal—which, perhaps, could have had a better chance of practical success than the more ambitious "positive" plan—found no official support on either side. Yet, neither the United States nor the U.S.S.R. could then (or can now) claim an urgent need for atomic industrial power. (Great Britain is the only major nation where this need may soon become acute.)

While the bold proposals of the Acheson-Lilienthal plan mired down in the morass of negotiations with the Soviet Union—distrustful of American purposes, evasive in her tactics, and fundamentally hostile to the idea of pooling the efforts of the Communist "world of the future" with those of the "doomed" non-Communist world—the atomic arms race gathered speed. Instead of being stopped, the minute hand on the atomic clock moved forward—in October, 1949, with the announcement of the first atomic explosion in the Soviet Union, and again in September, 1953, with the announcement of the first hydrogen bomb test in Siberia. (The

[3] For discussion of this plan, see the following articles in the *Bulletin of the Atomic Scientists:* Quincy Wright, "Draft for a Convention on Atomic Energy," I (April 1, 1946), 11–13; Cuthbert Daniel and Arthur H. Squires, "The International Control of Safe Atomic Energy," III (April, 1947), 111–13, 135; Cuthbert Daniel, "An International Moratorium on Atomic Energy for Power Uses" (letter to Sir Alexander Cadogan), IV (June, 1948), 183–84; D. Cavers, "Atomic Power versus World Security," III (October, 1947), 283–88.

[4] See Acheson-Lilienthal Report, *Bulletin of the Atomic Scientists,* I (April 1, 1946), 2–9.

first explosion of a full-scale American thermonuclear bomb was never officially announced!)

FALLING BEHIND TIME

Public understanding of the danger into which the atomic arms race had plunged the world has been slow in developing. In the fall of 1945, President Truman called the atomic bomb secret a "sacred trust" which America was going to keep "on behalf of mankind." In September, 1949, prompted by the Atomic Energy Commission, he announced the first atomic explosion in Soviet Russia; but at the end of his term, in 1952, Truman was still unconvinced that the Soviet Union really possessed the bomb! No wonder similar doubts persisted in the minds of many less well-informed Americans; most of them believed that if the Soviet Union had produced an atomic bomb it was only through stealing American secrets and that once all leaks were stopped she would rapidly fall behind in the race. This delusion still lingers on in the minds of many unfamiliar with the century-long Russian leadership in many fields of applied mechanics (including ballistics) and with the high present level of Russian nuclear physics.

Many military leaders, out of slowness of perception or because of vested interests in existing weapons, long refused to recognize the revolutionary nature of nuclear bombs. To prove his conviction that the atom bomb had been "oversold," a United States Air Force colonel offered to stand on one end of a runway at the National Airport in Washington while an atomic bomb was exploded at the other end. Navy men pointed to the survival of many ships after the Bikini tests in 1946 as proof that nuclear weapons did not make conventional naval forces obsolete. Much prodding, mainly by scientists (in particular in the "Project Lincoln" and "Project East River" reports[5]) was needed before American military leaders became convinced that the Soviet atomic-air progress made necessary extensive preparations for the defense of America's mainland—"active," or "continental," defense to detect and destroy bomb carriers, and "passive," or "civil," defense to minimize the losses inflicted by bombs which reach their targets.

In the last few years, the development of continental defense (radar chains and land-to-air guided missiles) has been considerably accelerated, but the progress of civil defense remains slow. Starved for men and funds and deprived of up-to-date information on weapon effects, American civil defense authorities have long con-

[5] For discussion of Project East River, see the special issue of the *Bulletin of the Atomic Scientists,* September, 1953.

tinued to toy with the idea of surviving atomic attacks as the British survived the blitz in 1940, by hiding in shelters. Many city buildings still display arrows marked "To Shelters"—which could as well read "To Deathtraps." The shock of the H-bomb explosion was needed for this suicidal idea to be replaced by the recommendation that cities in the path of an attack be evacuated.[6] Nearly two years later, no major American city has yet had the courage to try out the practicability of this concept in a full-scale exercise. As to dispersal of population and industry (which could afford a certain protection against all kinds of atomic attack), we as yet have almost none of it. Instead, new skyscrapers are going up in the most congested areas; and some basic industries have continued their concentration in metropolitan centers already replete with major targets.

Hesitation has also stunted our international policy in the atomic field. Hampered by "pre-atomic" (a better word may be "antediluvial") attitudes of influential segments of public and congressional opinion, our representatives entered the United Nations atomic energy control negotiations with no feeling of desperate urgency. They wanted to proceed "gradually"; but events have rapidly swept over their caution and their "safe" stages. With our allies, we have followed the same policy of *festina lente,* making only belated and restricted agreements on the sharing of atomic information, thus causing them to spend time and effort on investigations and tests, the results of which we could have predicted, and provoking a similar secretiveness on their part. This policy also hampers the training and integration of NATO forces into a single body of maximum efficiency.

Looking Backward

One way out of the dead end of the atomic arms race has been suggested by some whose eyes are habitually riveted to the past. If wars cannot be abolished, they say, let's make them "safe" again. Some military historians have been emphatic in denouncing the "perversion" of war which followed the development of long-range military aviation and reached its climax with the use of atomic weapons. Several speakers in a recent series of talks on "War and Society" presented by the British Broadcasting Corporation[7] argued that future wars must have limited objectives attainable by limited

[6] See "Planning an Effective Civilian Defense," and Val Peterson, "Mass Evacuation," *Bulletin of the Atomic Scientists,* X (September, 1945), 291–95.

[7] These talks were printed in *The Listener.* See also "In the Atomic Age Can War Be Limited?" by Raymond Aron, *Preuves,* December, 1955; Hans Speier, "War and Peace," *Bulletin of the Atomic Scientists,* XI (November, 1955), 346–49.

means and making defeat a painful but supportable blow to the loser. War will thus again qualify as a legitimate "continuation of policy by other means." In this way, they suggest, the destruction inflicted on the enemy, the mobilization of national forces, and the interference with the normal economic and political freedoms at home could be kept within reasonable limit, as they were in some "exemplary" campaigns of the eighteenth and nineteenth centuries. Of all the nostalgic dreams for the return of the good old days, this one seems to have the least relation to reality—and is also the one least worth dreaming!

As the size of atomic bombs became smaller, and the supply of atomic explosives grew far beyond all anticipation, some military planners conceived the idea of using atomic weapons mainly or exclusively against "legitimate military objectives" such as troops, supply dumps, or bridges. Some prominent atomic scientists were fascinated by this idea of "bringing war back to the battlefield." The report of the "Vista" project, at the California Institute of Technology in 1951, suggested that tactical use of atomic weapons may permit us to defend Europe without destroying her cities. Recently, however, Hanson Baldwin reported in the *New York Times* that large-scale army maneuvers in the South, in which about three hundred simulated "tactical" atomic weapons (from two to forty kilotons TNT equivalent) were used, showed that, if the bombs had been real, the main cities of several states would have been devastated by nearby blasts. Mr. Baldwin also commented that, in a real war, the side whose forward bases and stores were destroyed would have had no choice but to transfer operations to bases farther in the rear, thus necessitating the extension of atomic bombardment and the destruction of cities far behind the combat zone.

"Atoms Will Make Peace"

As the smug belief that atomic weapons had increased American security began to wane and hopes for atomic disarmament sank low, two new spurious optimistic philosophies began to spread. One can be related to the ancient belief that words, repeated often enough, have power over facts (a belief that was common in the age of spells and incantations and to which the success of modern advertising has given a new lease of life). Optimists of this type exult over "atoms for peace" and believe that the more we talk of them, the less the danger of "atoms for war" will become. The other optimistic philosophy makes virtue out of necessity and elevates the last, uncertain refuge of "deterrence through terror" to the dignity of a solid rock on which to base permanent national security. Optimists of this school worship at the shrine of "atomic stalemate" and see in the perpetuation of the capacity of two camps for instantaneous mutual destruction a guarantee of perpetual peace.

All scientists prefer to see science applied to constructive purposes; most of them favor the broadest possible international cooperation. The initiative of Lewis Strauss and President Eisenhower in arranging the Geneva Conference on Atomic Energy and in pushing forward the project of a UN-sponsored international atomic power agency has therefore found acclaim from scientists all over the world. There is, however, a wide distance between supporting these steps and believing that emphasizing the peaceful atom can become a serious deterrent to atomic war. This would be as irrational as believing that an international convention on the development of commercial aviation could become a guarantee against the use of planes for bombardment in a future war. The very slogan of the Geneva Conference, "Atoms for Peace," had an aura of advertising optimism. (Rationally—as was pointed out by General Phillips[8]—it should have been called "Atoms for Industry"; industrial atomic electricity can, and will, power defense factories as well as plants producing articles for civilian use, in the same way that atomic engines already propel submarines, even before they have started propelling commercial ships.)

Soviet propaganda has been quite willing to fall in with American semantic optimism. The Russian text of the "Atoms *for* Peace" posters, displayed all over Switzerland, asserted even more boldly that "Atoms *Bring* Peace." At the London conference on the "Perils of the Atomic Age,"[9] the leader of the Soviet delegation suggested that instead of considering this nasty subject, the conference would do better to talk about peaceful uses of atomic energy; this, he implied, was the way to prevent it from ever being used for destruction!

"Atomic Deadlock" Enthusiasts

The more "realistic" brand of optimism displayed by the enthusiasts of the "atomic balance of power" can be traced back to 1945, when the ultimate "saturation" of atomic armaments and the consequent inevitability of an atomic deadlock were much discussed among scientists. At that time Dr. Irving Langmuir, just back from an anniversary meeting of the Soviet Academy of Sciences, where he was strongly impressed by the quality of Russian science, argued

[8] Thomas R. Phillips, "The Atom in War," *Nation,* July 9, 1955.

[9] A conference of scientists convened in London, August 3–5, by the Association of Parliamentarians for World Government.

that the soon-to-be-expected development of Soviet atomic bombs was nothing to be worried about, since it would discourage warlike policies on the anti-Communist side, while American bombs would continue to restrain Soviet policies. Eight years later, the most imaginative of the political leaders of our time, Winston Churchill, similarly proclaimed that the possession of atomic and thermonuclear arms by both sides in a divided world can stabilize peace.

Since the revelation of the destructive power of the H-bomb, and of the radioactive infestation which the fission-fusion-fission superbomb can produce, the theory of peace through mutual deterrence has made many converts. These developments have made the decision to unleash a major war much more difficult, even for the most ruthless, but presumably still rational, aggressor. If one has to rely on terror weapons to keep peace, then obviously, the worse the terror, the better the chances of peace.

As long as the peace-through-deterrence idea is kept in proper perspective—as a desperate, temporary means to achieve a "breathing spell" during which mankind can search for a permanent solution, there is not much one can say against it; in fact, the last ten years have left us with little to rely upon as a safeguard of peace except deterrence. But when this concept is glorified as a final answer to the problem of war, a victory of technocracy over diplomacy, establishing the Strategic Air Command (and its Soviet equivalent) as permanent arbiters of peace in an unregenerate world, what an unrealistic utopia it becomes! What a fantastic perversion of the lofty concept of peace—a world of enshrined hatred and mistrust, in which the best efforts of each nation will be permanently bent toward improving its own machinery of mass destruction and outwitting other nations, which will be scheming to frustrate this machinery! What a degradation of science, to be put in the perpetual service of such an apparatus of terror, even if its purpose is to preserve peace!

The arguments put forward in 1945 against Langmuir's suggestions still stand, despite the advent of the hydrogen bomb and radioactive fall-out and the approaching advent of intercontinental missiles. The best that can be derived from such an atomic deadlock is a postponement of the explosion. That mankind can base its hopes for survival on indefinite maintenance of a perverse and immoral state in which evil is kept in check only by the fear of worse evil is a grotesque idea. Atomic deadlock can have positive value only if man uses the breathing spell it provides to develop a permanent foundation for peace without terror.

From Defense to Deterrence

The believers in peace through mutual terror imply that defense against the atomic air threat will never catch up with the attack. In this assumption they are probably correct. At present, the United States possesses a strategic air force supplied with hundreds, if not thousands, of atomic and thermonuclear bombs of various caliber and advance air bases permitting these planes to reach every spot on the globe. It is reasonable to expect that if vulnerability (or an uneasy political situation) should cause the United States to retire from these bases, longer-range bombing planes (and later, trans-continental missiles) will preserve American capacity to carry destruction to any spot in the world from bases in America. On the other hand, it is doubtful whether our present continental defense establishment could destroy, say, as many as one-half of the planes engaged in a mass attack on United States territory. It is even less likely that, whatever effort we may put into perfecting these defenses, we will be able to protect America so effectively that not even a small number of jet bombers will get through, lay waste many of our cities, and infect large areas with radioactive fall-out. This does not make continental defense efforts useless, for an effective continental and civil defense may still make the difference between a destruction that can be repaired and one that can only be avenged. Furthermore, without maximum protection of this type, our capacity for retaliation may itself become doubtful.

With the anticipated advent of intercontinental missiles streaking through the stratosphere at a speed of thousands of miles per hour, the relative effectiveness of defense is bound to be further reduced. Missile warheads may have to be lighter and, therefore, less powerful than plane-carried bombs, and many missiles are likely to go far astray; but, even an almost random rain of a thousand thermonuclear missiles could completely devastate large sections of industrial America. The same also applies, of course, to the Soviet Union, although its wider open spaces may make its position slightly more favorable. Satellite countries, as well as the free countries of Western Europe, are probably even now defenseless (and indefensible) against atomic air attack.

Barring entirely unforeseeable developments, a realistic long-range prognosis must thus assume for the future that major nations will maintain the ability to inflict, at will, enormous damage on their enemies and will have only a very limited capacity to protect themselves from a similar fate. This is an entirely new situation in history.

A strong state was presumed, in the past, to be able to shield its territory and its citizens from enemy attack. Fortresses, field fortifications, slit trenches, or tank barriers barred overland invasion, and armed ships guarded approaches by sea. The protection it afforded its citizens was, in fact, the first *raison d'être* of a state. If it failed in this task in the past, as it often did, it was the fault of its leaders—kings or statesmen who had failed to prepare for the emergency, generals who had used the wrong strategy or tactics, diplomats who had not secured allies. Today we still speak of "adequate national defense," conjuring the mental image of a country girded by walls and watchtowers (or their modern equivalents) and guarded by stouthearted men—a protection behind which national life can continue in safety. Yet, our military leaders know that they cannot provide this kind of national defense and, in all probability, never will be able to provide it in future. The word "defense" has therefore acquired a hollow sound. Instead of *defense,* we have—or soon will have—only *deterrence;* that is a tremendous change, and it may take more than a generation to comprehend it fully.

Security through Disarmament?

If military strength cannot physically protect a nation from destruction, perhaps diplomats can end the jeopardy by negotiating an atomic disarmament pact. This is, perhaps, the main hope by which mankind lives at present. And yet, two questions must be asked: First, do we really *want* such a pact now? We have to choose between the theory that the balance of two atomic threats stabilizes peace and the theory that a removal of these threats will make peace more secure. The Soviet Union has never wavered in proclaiming the second thesis; but Western policymakers have tried to support both at the same time! It would be, of course, a shock—and provoke much anti-American sentiment abroad—if we were to proclaim now that atomic disarmament, although possible and desirable between 1945 and 1948, has ceased to be so, now that all major nations have embarked on large-scale atomic production for industrial purposes, which will enable them to acquire atomic weapons very soon after the beginning of a war, even if they have destroyed their stockpiles during peace. Nevertheless, the United States should, perhaps, not hesitate to put a blunt statement to this effect before world opinion.

There is another and less controversial reason why negotiations on abolition of atomic weapons have lost conviction. Even if both negotiating parties were fully persuaded of the desirability of this measure, the belief in its technical possibility is dwindling. For several years now, the *Bulletin of the Atomic Scientists* has been pointing out how difficult, if not impossible, to solve are the problems involved in accounting for the increasingly decisive stockpiles through any kind of objective process of verification. In addition, the snowballing output of atomic materials for power purposes makes a reliable check on the diversion of this production from industrial to military applications more and more difficult; at the same time the immensely increased destructive potential of every pound of fissionable material (owing to the possibility of using it as detonator in a thermonuclear bomb) makes every undiscovered leak much more dangerous. Recently, these arguments have crept more and more often into public discussions of the problem. The Western negotiators seem to have recognized that unless the Soviet Union can show how the problems of past inventory and future leaks can be met, further discussion of atomic disarmament has little sense, except as propaganda. It would be desirable if scientists of neutral Asiatic countries—many of whom support the Soviet thesis that atomic weapons should be eliminated "under proper controls"—would produce some concrete ideas as to how this aim can be achieved under the conditions of "atomic plenty" (radically different from those of "atomic penury," under which the control idea was proposed in 1945 by American scientists).

Doubts about the desirability of the abolition of atomic weapons and the even graver doubts as to its technical possibility have turned the attention of Western representatives in the United Nations toward measures aiming not at abolition of these weapons but at the prevention, or early discovery, of preparations for aggression. We are pressing for the adoption of Eisenhower's "open sky" proposal; the Soviet Union has countered it with the proposal to station UN inspectors at ports and road junctions, an acceptable suggestion but obviously quite insufficient in itself. In fact, it may be doubted whether even aerial inspection will long remain a reliable method of discovering preparations for atomic air attack. At least, General Curtis LeMay, in a recent interview with the *U.S. News and World Report*,[10] said that after the bases have been properly dispersed and made self-sufficient in personnel and supplies a large-scale atomic air attack could be staged without obvious preparations.

To sum up, a realistic appraisal of the atomic disarmament situation leads to the conclusion that nations cannot reasonably expect deliverance from the nightmare of atomic destruction from their diplomats, any more than they can expect it from their defense chiefs.

This is what ten years of atomic energy have wrought. Truly, they

[10] December 9, 1955.

have shaken the foundations on which human society has rested since the dawn of history: the existence of tribes, nations, states, and empires, each a law unto itself, offering its members, in exchange for loyalty and obedience unto death, effective military or diplomatic security from overwhelming attack by a rival group. If history in our time were essentially that of the white race and the Western world—as was the history we were taught in schools—the import of the last ten years would be unambiguously clear. It seems likely that last spring's conference of the Big Four has signified the recognition of the truth by all of them, including the Soviet Union. If this surmise is true—and it is supported by the text of the recent Nehru-Bulganin communique[11]—the Soviet leaders have comprehended at last that atomic weapons have made a major war within the Western world suicidal for both sides. Not so long ago, Malenkov's assertion that "atomic war will bring about the end of all civilization" contributed to his downfall and was replaced by Molotov's smug, orthodox version that "atomic war will bring about the end of capitalism." Since then, the facts of Bikini atoll and Wrangel Island (if this is where the Soviet H-bomb tests are held) must have vindicated Malenkov. Technology has thus triumphed over dialectics; ten years of nuclear energy have proved more decisive than John Reed's "Ten Days That Shook the World."

Deadlock in West—Turmoil in East

World history is played in our time, however, on a stage much wider than Western Europe and America. The dominant event of the last ten years in the Western world may have been the advent of atomic and thermonuclear bombs; but in Asia and Africa, this factor is the awakening of native nationalism. While the West becomes immobilized by fear of atomic war, the East is in turmoil and flux. National pride and national hatred—passions which the Western world is about to outgrow under the common threat of atomic destruction —remain powerful forces in the young nations, first tasting (or yearning to taste) the heady wine of sovereignty.

It is this simultaneity of the end of history as we knew it in the Western world and the beginning of history in Asia and Africa that makes the present world situation so confused and unpredictable. The atomic deadlock in the divided West would in itself represent a highly dangerous situation, but, at least, the danger would be

[11] "The meeting of the heads of governments at Geneva in July, 1955, led to recognition by the great powers represented there of the futility of war which, owing to the development of nuclear and thermonuclear weapons, could only bring disaster to mankind." (From the text of the Indian-Soviet joint declaration issued on December 13, 1955.)

unique and clear to all, and the breathing spell provided by the deadlock could be long enough to heal the scar that runs across the map of Europe. The many-sided conflicts between the lusty, young national states of Asia and Africa and the temptations they offer for intervention by the two opposing camps add many insidious fire hazards to the one obvious danger in Europe.

In Europe, we are faced with the problem of maintaining peace despite a highly arbitrary bisection of the continent—cutting across the living bodies of nations—or subjecting the European nations entirely to unwanted domination, because we know that any attempt to change this division by force is likely to end in catastrophe. Asia and Africa are as yet far from a comparable state of rigid division. Sharp demarcation lines have been drawn across Korea and Vietnam and through the Formosa Strait, but even they can be made meaningless by events in neighboring states, Laos, Indonesia, Malaya, Burma, and Thailand, over which neither of the two camps has firm control. Until the Near East, Africa, and the Far East are all either clearly divided between associates of the two Western camps or—as most new nations fervently hope—consolidated into a separate stable political system, the peace of the world will remain precariously embroiled in the game of power politics played by twenty-five new sovereign nations, with the open or secret interference of master minds in Moscow, Peiping, Washington, London, and Paris.

The facts of the situation leave us, for the time being, no choice of policy in the West except to reinforce the political unity, military power, and economic stability of the Atlantic alliance, thus making still more rigid its deadlocked conflict with the Communist empire. Similarly, the facts of the situation leave us no other course in the Near and Far East but to exercise our ideological, political, and economic influence to assist the forces of political freedom and economic progress in the hope that, ultimately, these independent nations will find that their interests have more in common with ours than with those of totalitarian imperialism. However, the maximum success we can expect from this policy will be the extension of the rigid dividing line between the two camps, which now exists in the Western world, over the whole globe—a world-wide metastable equilibrium of uncertain permanency; and it is by no means certain that we will be able to achieve this degree of stabilization before sliding into war.

Such is the state of mankind after ten years of atomic energy—a precarious balance of mutual terror in Europe, a dangerous flux (from which we hope a similar rigid stabilization will ultimately emerge) outside it. If we reject the nostalgia for the limited wars of the past, if we cannot believe that atomic technology can, by itself, remove the jeopardy into which it has placed mankind (by emphasis on "atoms for peace," or by the dread of atomic destruction's keeping nations on the path of virtue—as the dread of hell is said to keep individuals from sinning), where, then, do we see a rationally possible and effective way out of the dilemma?

As scientists, we probably all—or almost all—agree that no solution can be based on the negation of facts, or a refusal to evaluate them as objectively as possible, and that much of the world's hopes are based today on such negation and refusal. Explaining and analyzing facts and educating public opinion to their acceptance, whether they are pleasant or not, has been one of the tasks of the *Bulletin of the Atomic Scientists* from its inception; and to this task it will remain dedicated.

However, facts alone do not suggest action unless a final aim has been set. This aim is, to our mind, the survival and progress of civilized society; and the analysis of the facts of the atomic age seems to show convincingly that this survival depends on the development of ethical standards at once broader and higher than practical men have been prepared to accept and live by—broader, in that they have to embrace all mankind and not just a single nation, and higher, in that they have to renounce selfishness and the resort to naked power, not only in relations between individuals within a nation, but also between nations. As has been often argued in the *Bulletin,* the first step toward this aim must be its clear recognition and adoption by our own country. This will put us on the road, starting from our real position in the world and leading in the right ultimate direction. From then on, we will have to find the path as we go along.

Is it wishful thinking, if it sometimes seems that despite new waves of nationalism, despite recurrent paroxysms of national and class hatred, a confused but powerful pressure for a new organization of mankind is rising throughout the world? In the West, people begin to recognize that their national states cannot give them the security for which they created them; in the East, newly awakened racial pride finds satisfaction in showing former colonial masters not only that Asiatics and Africans, too, can create powerful states but that, better than these former masters, they can keep peace and gain understanding among themselves. A ground swell is rising, sapping the emotional and ideological foundations of national sovereignties, changing the views and attitudes of men toward the state, and reversing their accustomed hierarchies of wants and values.

The atom did not bring about this change all by itself; it was foreshadowed by the sufferings and disillusionments which the two world wars brought to all participants, and particularly to those who had unleashed it. A comparison of the literature and art which came out of these two wars with the literature and art accumulated from the dawn of human history up to the twentieth century shows that the most sensitive section of society—its artists—had reassessed the formerly paramount values—national glory, military prowess, sacrifice on the field of honor—long before the advent of the atomic bomb; but it was this ultimate weapon that irrevocably devalued the old ideals in the eyes of whole peoples, including even their military leaders. Soviet speakers may accuse American military leaders of waiting for the chance to cut loose; but no American officer—probably no officer in *any* army in the world—now waits impatiently for a war to break out, as ambitious officers of all armies used to wait in the past, before the atomic bomb wrote finis to Napoleonic dreams.

Even before a new generation has succeeded the one that witnessed the breakthrough of mankind from the world of atoms and molecules into the world of atomic nuclei, humanity is in full upheaval. The forces of inertia and tradition, the spirit of "nothing is new under the sun," of "wars have always been with us and will always remain," are in world-wide conflict, in the minds of individuals and nations, with an inchoate, confused, but universal clamor for an end to history as a sequence of wars.

It is by no means certain that this vague and disorganized movement will assume clear form and triumph before the deluge of atomic war is upon us; but it is to the support and clarification of this aim that the *Bulletin of the Atomic Scientists* intends to devote itself in the second decade of its existence, as it has tried to serve it in the past.

12

The First Year of Deterrence

Last January, looking back upon the first ten years of the atomic era, I called them "Ten Years That Changed the World." With decisive changes in industry yet to come, changes in military technology brought about by the release of nuclear energy have already wrought havoc with the traditional meaning of the concepts that have dominated human history since its dawn: security in peace and victory in war. Deterrence by threat of atomic destruction delivered by air has largely replaced defense as a means of protecting ourselves and other nations and assuring freedom of action in world affairs. The inevitability of this revolutionary change impressed itself upon most thoughtful observers immediately after Hiroshima; but it was the year 1956, the eleventh year of the atomic era, that witnessed the first demonstration of its concrete meaning. Yet, the same year also saw a renewed burst of nationalistic passions and a display of selfish national policies, as well as much smug rationalization of the permanency of these traditional forces as the main movers in world relations. The contradiction between the living reality of the atomic age, calling for a harmoniously unified humanity, and the obsolete belief in the paramountcy of fractional interests has brought mankind a great stride closer to atomic self-destruction.

The year 1956 could be called A.D. 1—the first year of deterrence. Of course, as early as the Indochinese war and the Quemoy-Matsu crisis of 1954, American leaders started talking of massive retaliation as a means of preventing (or punishing) future aggression. This ominous threat, however, did not then become concrete and immediate. Similarly, the conviction that the peace in Europe has been and is maintained primarily by the American capacity for atomic destruction by air as the answer to any advance of Soviet military power beyond the 1945 demarcation line, though widely shared by military and political authorities, has never become alive in the imagination of peoples, to whom the word "war" still connotes primarily a vast movement of armies, as in the two world wars, and in whose minds the word "security" conjures up the picture of a protective wall of steel and concrete, manned by soldiers and armed with

From *Bulletin of the Atomic Scientists,* January, 1957.

guns. On the last occasion when war in Europe seemed to be a real possibility—the Berlin blockade of 1948–49—atomic retaliation was not openly threatened by the West, although its possibility must have played an important role in restraining the Soviet army from interference with the American airlift.

In the last year or two, it has become a generally accepted proposition that the threat of atomic destruction by air has ceased to be a unilateral American deterrent to aggression; but it has been widely assumed that world peace remains assured by the capacity of the United States and the U.S.S.R. to destroy each other by thermonuclear bombs. Despite the wide acceptance of the existence (or at least the imminent approach) of such a state of "absolute mutual deterrence," an air of abstraction and unreality still hung about the concept of "peace through mutual terror," until last November gave us a first object lesson in atomic air deterrence. Ironically, it was not the original protagonist of this concept, the U.S. Strategic Air Command (SAC), who gave this demonstration but that newcomer in the field of atomic air strategy, the Soviet High Command. The opportunity was supplied by the discord in the Western camp, which led to the separate Anglo-French excursion into Egypt. Suddenly, the Soviet Union found herself in the position America had enjoyed between 1945 and 1950: able to threaten atomic destruction by air without fear of immediate retaliation in kind. Furthermore, in contrast to America's situation after 1945, the Soviet Union was not confronted with powerful conventional armies poised on the opposite side of the East-West demarcation line. With the capacity for immediate and ruthless utilization of a strategic advantage, characteristic of a totalitarian regime, the Soviet leaders seized upon the opportunity. A few days after the Anglo-American ultimatum to Egypt, Britain and France were presented with a virtual Soviet ultimatum threatening, by clear implication, atomic destruction by air of both countries if they did not call off the Suez expedition. This threat may have been a bluff—at least, in part. It is by no means certain, for example, that (as Bulganin's letter of November 5 implied), the Soviet Union already possesses rocket missiles able to deliver hydrogen warheads to targets in Western Europe. General Alfred M. Gruenther expressed doubts on this point; it may be significant that the reference to rocket delivery, contained in the text read over the Moscow radio, disappeared from the note presented a day later to Prime Minister Anthony Eden. However, this point is not essential, since delivery of H-bombs by manned planes, of which the Soviet Union is undoubtedly capable, would have been quite sufficient to lend weight to the threat. In any case, by the time the note was officially delivered, the ultimatum had already achieved its aim: England and France had announced their readiness to stop military operations and to withdraw their troops from Egypt as soon as the face-saving United Nations police force could replace them.

For the first time in history, a major world power (or rather, two such powers) were stopped dead in their tracks at the outset of a victorious operation which they considered absolutely vital for their survival. The Suez expedition had appeared so important to England and France that they considered a rift with their principal ally, the United States, as well as unanimous reprobation by the United Nations, and even a possible breakup of the British Commonwealth, as not too great a price to pay for it; yet, the threat by the Soviet Union to turn on them the fury of Russian "deterrent" weapons was enough to cause the abandonment of the whole enterprise.[1]

This demonstration of the power of atomic air deterrence is a turning point in history, and we should not be distracted from recognizing its significance by incidental events, such as the diplomatic pressure the United States exerted upon England and France or the speeches given and resolutions passed in the United Nations. These latter consequences the leaders of England and France had probably anticipated and discounted. They hoped—and probably with reason—that rapid military success of the Suez occupation and the downfall of the Nasser regime in Cairo would be followed, as similar successes achieved by force had been in the past, by a gradual acceptance of the *fait accompli* by the whole world. What England and France did not anticipate was a readiness of the Soviet leaders to unleash all-out atomic war in response to a local conflict so far from their own borders.

Of course, after the decision to stop the Egyptian invasion was taken, all the previously dismissed arguments against it—disunity at home (particularly in England), moral pressure of the United Nations, threat of oil shortage and the consequent possibility of an economic squeeze by the United States—have acquired new strength. They have helped to push England and France toward complete and unconditional abandonment of all positions of strength they have acquired in the Suez area.

[1] As C. Douglas Dillon, U.S. Ambassador to France, said in a radio interview on C.B.S. on December 11, 1956: "I don't think it was moral suasion [that has caused France and Britain to abandon the Suez invasion], they went ahead with their Suez operation despite sharp criticism in the United Nations and the United States. Nor was it the pressure of dwindling oil supplies—that hadn't had time to make itself felt. . . . The only new element that had come in was these Soviet threats, which were very, very strongly phrased."

Now, a myth is beginning to grow. It is so much less alarming for England to have yielded to the voice of the majority in the United Nations than to have been compelled to change policy by a brutal threat of Soviet H-bombs! It is so much more satisfactory for the United States to have asserted the power of its leadership in the fight against colonialism than to have merely ridden the coattails of the Soviet might! The affair can even be made to appear, in retrospect, as a triumph of virtue—virtue of moral integrity on the side of America and virtue of contrition on the side of England and France! And the U.N. can be credited with having demonstrated, for the first time, its capacity to stop aggression by major powers. Such is, in fact, the rapidly growing myth of the Suez fiasco, abetted, for different reasons, by practically every national leadership in the world—American, Anglo-French, neutralist, and even Soviet.

The first lesson of the November events seems to be that if Western Europe wants to regain the freedom of action the United States and the Soviet Union alone now seem to possess on the world stage it must, first of all, accumulate its own arsenal of atomic and thermonuclear weapons. However, even if Western Europe, overcoming its traditional political disunity, could make the effort needed for this purpose, it will remain at a disadvantage in the effectiveness of its deterrent power, because geography puts the small Western European area within easy range of all kinds of Soviet bomb carriers, whereas many Soviet targets are far from the airfields and missile-launching sites in Western Europe.

In the absence of its own atomic air power, Western Europe is and will remain dependent upon the United States to match the Soviet threat. It was because the American attitude toward the Soviet threat was ominously vague that England and France found themselves unable to resist on November 5. The American answer, reasserting the mutuality of deterrence, eventually did come—but only several days after the Anglo-French expedition had been stopped dead at Port Said. This answer was contained in the farewell speech by General Alfred M. Gruenther to the NATO command on November 13 (undoubtedly approved beforehand in Washington). General Gruenther said that any military incursion of the Soviet Union into Western Europe would lead to instant destruction of the Soviet Union by the American air force, and he warned that even an elimination of African or European SAC bases by Soviet rockets could not prevent the retaliatory attack, since it could be delivered by long-range jet bombers from bases in America. Undoubtedly, it was to emphasize the latter point that it was officially revealed on November 27 that eight B-52 jet bombers had completed non-stop flights over circuits fourteen thousand to seventeen thousand miles long, between Maine and California via Greenland, the North Pole, and Alaska.

This American counterthreat has restored the balance of atomic air deterrence in Europe, and the East-West "dialogue" over the Middle East has switched to more conventional topics, such as alleged Soviet arms delivery to Syria and the possibility of an intervention by Soviet "volunteers" in Egypt. This does not prove, however, that, to the extent that the West can match the air-atomic destructive power of the Soviet East, it enjoys—as of old—a full sovereign freedom of action in pursuit of its interests on the world stage. The paralysis of the West before the Soviet suppression of the Hungarian revolution is a clear warning of how restricted this freedom actually is. In contrast to Anglo-French indecision in the face of Germany's attack on Czechoslovakia in 1938, the present paralysis represents more than a lack of political will; rather, it follows from an actual, physical incapacity to devise and carry out an adequate and purposeful action. Even in full possession of their power, including the power for atomic destruction, the Western powers cannot play all their cards in today's power game, except when their national interests are so overwhelmingly involved that the threat to use the ace in their pack, atomic air power, becomes justifiable at home and plausible abroad. An American threat to use atomic weapons in defense of a cause such as the liberation of Hungary would not have found full support at home and could have been taken for a bluff in Moscow. It would therefore have run the risk of ending either in painful retreat or in actual atomic holocaust, thus failing in its primary—and only reasonable—purpose: deterrence.

In situations of this type, totalitarian regimes have a much greater freedom than democracies. Such a regime can declare that a conflict anywhere in the world is of such importance that it justifies the threat of atomic war (as the Soviet Union did in the Suez case). It does not need for this a genuine popular support at home; and its threat cannot be dismissed abroad as a bluff, because it comes from a regime rightly held capable of brutal and irresponsible action. Mr. Dulles may have invented the game of brinkmanship, but Mr. Bulganin can play this game with much greater ease!

The deterrence dialogue between the East and the West did not quite end with General Gruenther's judiciously delayed answer to Bulganin's ultimatum. The Soviet premier had one more word to put in, and he did it in his disarmament letter to President Eisenhower on November 17. Unfortunately, the American press, which has largely missed the full significance of the Soviet ultimatum to Britain and France (despite its incredible bluntness) and has given little prominence to General Gruenther's equally blunt answer, has altogether neglected this important passage in Bulganin's letter, probably dismissing it as part of the vociferous propaganda which, since Andrei Vishinsky's time, has customarily preceded every new Soviet proposal in the disarmament area.

In the passage in question, Bulganin suggested, albeit obliquely, that America should not believe that the mutual threat of atomic destruction by air restricts the freedom of the Soviet armies to invade Western Europe. He proclaimed the Soviet capacity to conquer Western Europe "even without the use of up-to-date nuclear and rocket weapons," saying that "the strategic situation as it has developed itself in Western Europe now is even more favorable for the armed forces of the Soviet Union than [it was] at the end of the Second World War." What strategic situation does he have in mind? Obviously, the situation of mutual atomic air deterrence. It is easy to see why, in Bulganin's opinion, the strategic situation is now more favorable for the Soviet Union than it was immediately after the war: unilateral American capacity for atomic air attack did balance, at that time, the Soviet preponderance in conventional power; whereas now this capacity is canceled out by a similar capacity of the Soviet Union. Bulganin seems to believe that America will be unwilling to risk her own atomic destruction by unleashing an all-out atomic war in defense of Western Europe. Consequently, it is only the Soviet Union's "love of peace" and not the threat of United States air power that prevents the Soviet Army from taking over Western Europe; and, Bulganin implies, one day this love of peace may be put to too severe a test. In this way, the American resolve to answer a direct Soviet attack on Western Europe by mass retaliation against the Soviet Union, so categorically reasserted by General Gruenther, is questioned by the Soviet leader; and by the very token of this doubt, our threat of retaliation loses some of its deterrent power.

The obvious second lesson of the situation, so gleefully presented by Bulganin, is the necessity for the West to build up with all speed the unfinished NATO army in Europe, until it appears capable of stopping the advance of Soviet tank divisions without resorting to all-out thermonuclear war; only then can this force adequately supplement the deterrent power of our atomic air arm and prevent a fateful miscalculation by the Soviet leaders. In an interview given by General Lauris Norstad to the *U.S. News and World Report* on November 30, the new commander of the NATO forces did not hide the fact that the present forces under his command are insufficient for this purpose. There is, however, no reason for this situation to be permanent. The Western alliance has a considerable preponderance in manpower over the Soviet Union—not the other way around, as is so often taken for granted in disregard of statistical facts. (The countries of Eastern Europe obviously do not effectively add to Soviet manpower and are not likely to do so in the near future. The structure of the Soviet economy has now become similar to that of the West, and its former advantage—a large peasant population which could be put into uniform without endangering industrial production—has all but disappeared.)

In the meantime, all must be done to dispel any doubt the Soviet leaders may entertain of the certainty of an all-out atomic answer by the United States to an invasion of Western Europe. There is one way to achieve this aim: not merely to restore the shaken political unity of the West, but to make it much stronger and more permanent than it has been in the past. A decisive change in this direction can be achieved by the conversion of the NATO alliance into a political body so thoroughly integrated that no doubt could arise that an attack on any part of it would provoke the same reaction as an attack on the whole of it. It may be difficult to achieve this higher unity immediately in the wake of a sharp rift between the allies (although, given imagination and will, even this is not impossible); but at least the aim of political integration of the West must be acknowledged by leaders in America and Europe, replacing in their minds the aims of an old-fashioned military alliance, which are inevitably temporary and exposed to all the divisive forces of the independent—and often contradictory—national policies of the alliance members. The December NATO conference has brought signs of increased recognition of this aim by Europeans but a sharp rebuke of it by Secretary Dulles.

The danger of war in Europe has become much more urgent—and less apt to be kept in check entirely by mutual atomic air deterrence—since the sharp dividing line between the two power spheres, from the Baltic to the Mediterranean, has shown signs of instability. (The whole concept of "peace through mutual deterrence" is built on the premise of a rigid and permanent division of the two camps.) The heroic—almost successful and still not quite abandoned—attempt of the Hungarian nation to break away from the Communist camp has revealed deep cracks in the supposedly monolithic structure of the Soviet empire. The fight of the Hungarians can be followed, at any time, by a second upheaval in Poland, an uprising in Eastern Germany, or a revolution in the Balkans.

The inevitable—and increasingly desperate—attempt of the Soviet leadership to keep their empire from disintegrating can lead to situations of gravest danger for peace, situations in which the Iron Curtain may cease to be a mutually respected rigid barrier separating the "legitimate" (except for verbal protests) field of operation of Soviet tank divisions from the now "forbidden" West. If such a situation should arise, it will be of crucial importance for our future, if not for the survival of all mankind, that the West be able to interfere by means other than the threat of unleashing an all-out thermonuclear destruction.

Here lies, at present, the most terrible danger to mankind. If American political leadership can be blamed for its handling of the Middle Eastern situation, it is because it has assigned to the necessity (extremely urgent) for America to acquire the support of Asiatic and African nations a higher priority than to the necessity (even more urgent) of preserving the unity of the West. It has thus contributed to the creation of a danger that it has been the consistent primary objective of American foreign policy to avoid, the danger of a new war in Europe. The third lesson, therefore, of the two closing months of 1956 is the necessity to restore and reinforce the unity of the West, giving to this task the paramount priority among all American political aims.

The lessons taught by the crisis of November, 1956, if taken to heart, can provide only temporary help in an emergency. They can help in the establishment of a "pseudo peace," based on mutual deterrence in the most rigid form—a state of affairs in which a move by either of the two camps anywhere along their momentary frontier will be restrained by a manifest danger of unleashing both a major conventional war (fought with atomic weapons) and a full-scale thermonuclear catastrophe. However badly we need to achieve this metastable state, only a few technological dreamers, with little feeling for human reality, can look forward to it as a truly stable and desirable final solution. Mankind will not advance beyond the doubtful ideal of a frozen state of ubiquitous mutual deterrence unless it derives, from the events of 1956, a deeper and more stirring lesson: the recognition in these events of the demise of the whole concept of "freedom of action" on the world stage of all "sovereign" nations pursuing their "national interests."

It is doubtful that this lesson is being properly read and taken to heart at the present time. In the most stirring event of 1956, the freedom fight of the Hungarian people, world opinion was aroused, above all, by the demonstration of the will of a downtrodden people for national independence. Yet, how obsolete and oddly unreal now is the ideal of absolute national independence! What was originally a source of strength in the movement for human freedom in the Western world—the association between the ideas of individual freedom and national independence—has now become its fatal weakness. These two ideas have been inextricably entwined in the hearts and minds of men ever since the fight for American independence and, particularly, since the national liberal uprising in Europe in 1848 and the many subsequent struggles when men sought to free themselves from the oppressive European empires. It is this fusion of the ideals of personal freedom and national independence that has prevented the democratic nations of the West from finding at the summit of their power—after the First World War—the path to a truly effective democratic world organization, which alone could have put an end to wars and thus preserve and encourage the spread of individual freedoms. The struggle for national self-determination, begun at Lexington in 1776 and elevated by President Wilson in 1917 to a sacred ideal of democratic humanity, still continues in 1956, in Europe as well as in Asia and Africa, although absolute national independence has become obsolete, if not impossible, in consequence of the world-wide progress of technology and the global integration of national economies engendered by this progress. The achievement and preservation of individual liberties is now clearly tied up with the integration of mankind into a single organic body and effective abolition of the power of nations to make war—the ultimate expression of national sovereignty. Yet the hearts and spirits of men, not only in newly emerging nations of Asia and Africa but even in the history-wise peoples of Europe, still respond most strongly to appeals for national liberation and independence. "Non-interference in internal affairs of a sovereign nation" remains a hallowed principle to which everybody feels obliged to pay at least lip service, although the examples of modern totalitarianism in Russia and in Germany should have made it abundantly clear to all who are not blinded by tradition that there are internal affairs of nations which are of vital, and therefore legitimate, concern to all mankind. No nation can remain free and secure in our technological age if a political group is permitted to destroy individual liberty and establish totalitarian domination over any other nation, because under the cover of national sovereignty this domination can be made the basis for a threat of destruction of all other nations.

The fight against Soviet tyranny has assumed more acute forms in Poland and Hungary than in Czechoslovakia and Bulgaria, because in the first two nations the traditional hatred of Russian domination has remained alive even within their Communist parties—not because Hungarians and Poles were more attached to individual freedom than Czechs or Bulgarians. (As a matter of fact, the Czechs, in particular, have had a much better record of securing and enjoying political freedom than the Poles and Hungarians, who have preserved a feudal social system longer than any other nation in Eastern Europe.)

It has been reported that some Hungarian rebels have proclaimed that they were fighting not for Hungarian independence but for the integration of Hungary with the West; but, undoubtedly, the mainspring of the uprising has been, and remains, the striving for national independence. This is fully understandable; nobody can criticize the heroes of Budapest if in their minds political tyranny appears synonymous with foreign, Russian domination. It was pathetic, however, to see the representatives of democratic peoples of Europe, Asia, and America stand up in the United Nations and denounce the suppression of the Hungarian revolution as, above all, "interference with the internal affairs of an independent country." (In a grotesque counterpart, the Soviet and satellite speakers insisted that the United Nations has no right to concern itself with the "internal affairs" of the "independent" Hungarian people.) As if the most tragic aspect of the Hungarian drama was that the weapons that crushed the fighters for freedom were Russian and not Hungarian weapons! As if world compassion becomes legitimate only when tanks shoot up students, workers, women, and children who speak a different language and belong to a different race than the tank crews! This emphasis on the rights of *nations* and on the interests of national *states* in an age which calls, above all, for respect for the rights of the *individual* and recognition of the interests of *humanity* is expressed in 1956—at least, officially—by Communists and anti-Communists, by former imperial and former colonial peoples, alike; and this bodes ill for the future of mankind.

The world-wide yearning for atomic disarmament is fed, in large part, by a related reluctance of nations to recognize the irrevocable change which the discovery of atomic energy has brought into their situations and a nostalgic desire to return to a time when nations enjoyed sovereign freedom of decision, in war and peace, and could play the game of power politics without fear of their own and universal destruction. If, by some miracle, atomic weapons could be made to disappear, history could return into its old groove; and this miracle seems to be the fervent hope of all national governments in the world, from Moscow to London and from New Delhi to Washington. Yet it is a completely and utterly unrealistic hope. The expansion of technology and the intertwining of national economies have already produced a *de facto* unified world in which any major conflict means the opening of a wound in a living body and any major war entails a bloodletting from this fissure which weakens the organism as a whole; but prior to the discovery of atomic energy, mankind could hope to pass through a sequence of such bloodlettings and cicatrize its wounds after each of them. The great jump from the chemical into the nuclear age has made the continuation of this historical "tradition" impossible. This jump is irreversible: Never again can there arise a system of independent and sovereign nations, able to conduct their international policies as their "national interests" dictate them, resorting to force when other methods fail, without endangering not only their own existence but the survival of the whole human race.

This does not mean that we do not need to strive for atomic disarmament but that we can strive for it purposively only in the context of a general policy directed toward the integration of mankind into a single community and not in a groping, backward way in the vain hope of restoring the lost freedom of action to individual, sovereign nations. Unfortunately, it is in the latter mood that most nations in the world (including the so-called neutral ones) now seek atomic disarmament—and will never be able to find it. Progress toward disarmament and security is now possible only after unreserved recognition of the actually existing, and from now on forever inescapable, community of fate of all mankind and a consequent readiness of individual nations to subordinate national interests to the interests of mankind as a whole. There is no national government now that does not bear responsibility for all humanity and not merely for its own people.

The question of the continuation of thermonuclear bomb tests, so violently discussed in 1956, is of relevance in this context. The arguments used in this controversy dealt, on the one hand, with the importance of the tests for American national defense and, on the other hand, with their danger to the health of the present generation and the hereditary endowment of its descendants. Whether this danger is great or small, the most important thing is that it is universal and compulsory. The American government (or any other government conducting such tests) is—or should be—weighing the national interest of its own country against the danger to which the pursuit of this interest exposes all men on earth. It is traditional for a nation to accept hardships and risks to defend its national independence and to further what it considers its national interests;

but can a nation—any nation—claim the right to impose, without consultation, hardships and risks on all other nations? Not unless its paramount aim is the common interest of all mankind; not unless it accepts its responsibility toward all men and not only toward its own people! It is not enough to say that in perfecting her atomic weapons America actually serves the interests of all nations, because Communist tyranny threatens all men (including those who are unaware of this threat). As long as our overriding concern is with our own nation only, we cannot expect mankind to accept our judgment in questions affecting all of it. What is needed is a genuine change of heart and a convincing demonstration that the common interests of mankind, rather than the limited interests of a fragment of it, are our prime concern. (Of course, the same applies to all other governments, too; but the failure of others to acknowledge their share of responsibility does not make ours less compelling or urgent.)

It may be asked: Is not the proper answer provided by the policy of "supporting the United Nations" and "acting through the United Nations," which the United States has pursued in the Suez crisis? Not as long as our support is restricted to policies which happen to coincide with what our leaders consider our national interest! Like every other nation, we want other nations to obey the will of the United Nations (as expressed in resolutions for which we have voted!) even if these desires run contrary to the national interests of the country to which they are directed. The acid test comes when the desire of the majority of nations contradicts our own national interests. So far, every nation placed in this position has failed the test and has put its national interests above considerations for the community of nations. The Soviet Union has flaunted the desires of the UN majority on innumerable occasions; so has France when the Algerian question came up; so has India in the Kashmir dispute; South Africa is doing so in the question of racial discrimination. England and France were all set to defy the United Nations in Egypt and were only restrained from doing so by Soviet threat of violence. Can the United States honestly assert that our government would have acted differently from other countries if an important American national interest were involved—as, for example, in the Guatemalan crisis or in the question of H-bomb tests (which may well come before the forum of the United Nations sooner or later)?

The even more relevant question at the present time is: How would the United States have reacted if the Republic of Panama had seized the Panama Canal (placing us in a situation similar to that of England and France in Suez) or if other nations had claimed the right to participate in the supervision of this canal (thus placing us in a position similar to that of Egypt)? The answer is clear; we would have acted as we saw fit according to our national interest and resisted "interference in our internal affairs." Yet, there is probably only one way in which America can effectively strive for an equitable solution of the Suez controversy, and this is by proclaiming that no waterway of vital importance to many nations can be treated as private property by the one of them whose territory it happens to traverse. In the light of this principle, it would become obvious nonsense to consider the desire of Egypt to exercise sole authority over the Suez Canal as part of the fight for liberation of former colonial countries from imperialist oppression. The Suez canal is not a natural resource of Egypt that has been appropriated and exploited by colonial powers. It is not a natural resource at all but an artifact which European technology has created; it serves the interests of world trade and thus of all nations. It is not colonialism to insist that this enterprise should be administered by those who use it and its income used primarily for its own maintenance and improvement. This seems to be no more than obvious justice; but in a world where national possession is the supreme law and national sovereignty the supreme good, this justice can find little recognition. For this, the supremacy of national sovereignty and national interests in situations involving wider groups of people must first be challenged as a matter of general principle. If the United States would announce such a principle (as President Truman once proposed and Senator Ralph Flanders, of Vermont, among others, has again suggested recently), if we would express our willingness to apply it also to the Panama Canal, a new moral foundation would be laid for a just solution, not only to the Suez controversy but to many future controversies. Of course, there is a long path from the proclamation of such a principle to the elaboration of an equitable procedure for its application to specific cases, each inevitably different from the other (as the Panama Canal situation is different—legally, technologically, and economically—from that of the Suez Canal); but a break in principle with the recognition of unrestricted supremacy of national interests in areas where many nations, or all mankind, have a legitimate interest is necessary if mankind is to advance toward a stable, peaceful future—and it has no other!

These are some of the lessons of A.D. 1—the year of the Hungarian revolution, of the Suez drama, of the H-bomb test controversy; the year that has seen the first prototypes of intercontinental missiles raising their heads to the sky on their giant scaffolds and has heard world powers threatening to wipe each other off the map at the press of a button, knowing that this is not an empty boast—or, at least, will not long remain one.

13

New Year's Thoughts—1958

Introduction

Five years after the following "thoughts" were written down, disarmament still remains the fond hope of mankind and the official aim of all statesmen. The Soviet government has made it even more attractive by specifying "complete and universal" disarmament. The United Nations has unanimously endorsed this formula—voting against it would be like voting against virtue! Despite this agreement in principle, disarmament negotiations, including those on the modest first step of a nuclear test ban, continue according to the prescription: "one step forward, one step backward," a day of hope, a day of gloom. Despite these frustations, the search for a disarmament formula, acceptable to both sides in the cold war, goes on, officially in Geneva and unofficially in private groups, such as the Conferences on Science and World Affairs (COSWA). This persistence deserves all possible support, despite the pessimistic conclusions reached on New Year's Eve, 1957, and often repeated by the author since. I may be wrong in seeing insurmountable technical difficulties as blocking controlled disarmament in the era of nuclear plenty and of underground rocket launchers.

The conception has gained wide acceptance in recent years that the true difficulty of disarmament is political. Technical controversies, it is said, are only pretexts to avoid politically undesirable agreements. For political reasons, neither of the two sides wants to assign disarmament the high priority in their set of political aims, without which it will never be attained. If a proper revision of political priorities could be achieved, control problems would become soluble. The technically impossible requirement of absolutely reliable control systems by the West, as well as the frustrating reluctance of the East to accept any controls that would reveal something about its military capabilities and dispositions, would become negotiable. There is some truth in this view. The West is becoming educated to the realization that *absolutely* safe disarmament controls are impossible, that a certain risk of clandestine activities cannot be entirely avoided, but that this risk can be preferable to the risk of an unwanted and unplanned nuclear war, inherent in an uncontrolled arms race. The East, at the same time, is being educated by the development of various long-range physical detection methods, such as "spy satellites," that the asset of "security by secrecy" is waning and that it may be better to place this asset on the bargaining list before it becomes too badly depreciated.

However, I believe that the disarmament equation will remain unsoluble as long as no systematic attention is paid to the factor of trust.

The less acute the distrust between nations, the less vociferous will be the clamor for the impossible 100 per cent safe controls on the one side and the less stubborn the reluctance on the other side to accept some controls affecting the secrecy of its military posture. I believe that trying to achieve disarmament without pursuing simultaneously, by all possible means, an increase in mutual trust is like hitching the horse behind the cart; one can move forward in this way, but with what difficulties! A widespread view, shared by many Soviet speakers, is that trust will be the product of disarmament—the further the latter advances, the greater it will become. But this point of view leads to a vicious circle which can be broken only by decisive measures, leading to a reduction of distrust between the East and the West, in areas other than the arms race. I have in mind agreements of two kinds: standstill agreements in areas of acute political conflict (exemplified by the Laos settlement) and agreement on active co-operation in non-political areas (exemplified by the International Geophysical Year).

In his recent interview with William Randolph Hearst, Mr. Khrushchev said that the Soviet Union already has operational intercontinental missiles with nuclear warheads, capable of hitting targets in America. Whether one accepts this statement literally or thinks that Mr. Khrushchev, in his exuberance, may be running a little ahead of the facts makes little difference. In any case, the statement shows that we have reached a point of no return. For one or two years, thoughtful observers have warned of its approach. Colonel Richard Leghorn urged that we had but a few months left to stop the arms race.[1] Once a nation, he said, has developed ICBM's with nuclear war heads and built concealed, well-dispersed underground bases to launch them, reliable international control of this "ultimate weapon" will have become technically impossible. Under these conditions, other nations will not be willing to forego acquiring similar weapons. The era of permanent mutual terror will be ushered in. Attempts to stop or reverse the arms race will become futile, unless mankind first creates a climate of co-operation and trust or establishes a world organization with effective police power—and the second is hardly feasible without the first. In the absence of a radical change in international relations, the arms race will go on and on.

The warning, made by Colonel Leghorn, that the arms race must be stopped now or never did not sink in. The politicians had no sense of extreme urgency. Since 1945, the development of their ideas

From *Bulletin of the Atomic Scientists,* January, 1958.

[1] Richard Leghorn, "Controlling the Nuclear Threat in the Second Atomic Decade," *Bulletin of the Atomic Scientists,* XII (June, 1956) 189–95.

has been consistently behind the progress of weapons technology. Of course, the control of space vehicles was included in the American proposals to the Disarmament Committee; but Mr. Harold Stassen did not press for its immediate consideration. Now it is all over. Controlled disarmament, in which the whole world has been putting its hopes for the last twelve years, has ceased to be a rational possibility—at least, as long as mankind remains divided into mutually distrustful sections. For the time being, there are only two alternatives left: disarmament without effective controls, or acceptance of the state of "mutual deterrence" as a guarantee of peace, until a true world community has been created.

Let us look at this tragic situation with open eyes and stop trying to fool ourselves and the world. Mr. Khrushchev may again proclaim his readiness to turn all rockets and atomic explosives to non-military uses. This is mere talk. Even if these were his true intentions, it is not in his power to carry them out. The Soviet Union has no way of convincing the world that it has dismantled its nuclear ICBM bases; and America will have no way of persuading the Soviet Union that it has done the same. It is time to reconsider our plans in the light of this fact. Disarmament negotiations have been a make-believe for the last several years, because the will to disarm was too weak on both sides and all proposals were hedged with conditions intended to secure military advantages for one or the other side. All that these negotiations accomplished was to keep up the hopes of anxious people, hopes which now have become objectively futile. Will our political leaders realize this? Will they find the courage to tell the bitter truth to the American people, to the world? Will they start looking fervently for a new approach to the problem of permanent peace?

The leading article in the *Bulletin of the Atomic Scientists* a year ago was entitled "The First Year of Deterrence." It reviewed the events of 1956, in particular the Suez expedition and the Hungarian uprising, and concluded that the outcome of both crises showed that the world had entered the era of mutual deterrence. Every attempt of a nation—even a major nation—to exercise its accustomed "freedom of action" on the international scene, by the use of force, could now be frustrated by the threat of nuclear destruction at the hands of one of the nuclear powers. This is what happened at Suez, where two major Western nations, following what they fully believed were their legitimate national interests, were turned back by the Soviet threat of nuclear bombardment. This interpretation, highly inconvenient to all sovereign nations, was not the one most widely accepted. More popular interpretation credited the fiasco to American indignation over the Anglo-French aggression, Liberal and Labour opposition in England, the danger of the intervention of Soviet "volunteers" in the war in the Near East, or the rising power of Arab nationalism. All these factors did contribute to the Anglo-French withdrawal, but the one most important—and unprecedented—factor was the existence of atomic air power and the apparent readiness of the Soviet Union to use it. Recently, Marquis Childs wrote:

> Soviet Prime Minister Nikolai Bulganin sent notes to France and Britain, threatening that if the attack on Egypt did not cease, rockets would begin to rain down on Paris and London. Publicly, the West at that time discounted the rocket threat, claiming that the moral persuasion of United Nations' action had succeeded in stopping the Egyptian invasion.
>
> Actually, the Bulganin note produced something like panic in the Western capitals. The rocket threat may not at that time have been taken with complete seriousness, although since then the existence of launching sites in Western Russia capable of sending missiles with a range sufficient to hit all points in Europe has been verified. It was undeniably a factor, however, and perhaps a major factor, in ending the Suez conflict.[2]

Some consequences of the state of mutual deterrence were analyzed in an article, "The Frozen Map," in the June, 1957 issue of the *Bulletin*. It was argued that the existence of this state makes it practically impossible to change by force even the most unjustifiable boundaries of nations or spheres of influence and that our political planning should be revised in the light of this fact. If a possibility exists that the most dangerous frontier in the world—that dividing Germany—can be wiped out just before its ultimate freezing, by the elimination of foreign troops from both parts of Germany, this possibility should be explored. Mr. George Kennan has eloquently presented a similar view in one of his recent lectures. More generally, the approach to the state of mutual deterrence may permit a disengagement of the opponents in all areas where their close contact creates the most immediate danger of an outbreak of hostilities.

The year 1957 was the second year of deterrence, and again deterrence has worked. The attempt by the United States to assist anti-Soviet elements in taking over the government in Syria, a maneuver quite legitimate in the established tradition of power politics, failed because of Soviet deterrence. The Soviet attempt to eliminate Turkey as a barrier to the expansion of its influence into the Near East was similarly frustrated by a threat of American deterrent power. For a few critical days, it seemed that the Soviet leadership

[2] *Chicago Sun-Times,* October 19, 1957.

might be tempted to disregard the American warning, on the assumption that it was a bluff, that America would not expose its own cities to destruction to stop the Soviet armies from invading Turkey. However, caution prevailed on both sides, and the crisis passed. The incident showed, however, the precariousness of the state of mutual deterrence, idealized by some military scientists as a secure state of stable peace.[3] Its maintenance depends on the belief of each side that the other is prepared to go through with the threat, whatever the frightening consequences may be. One day somebody will doubt it and make an error. Throughout the age of deterrence, the fate of mankind will depend on the counsels of caution prevailing within the governments of all major nations, however irresponsible, power-drunk, or even mad their rulers may be.

The development by the Soviet Union of intercontinental nuclear rockets also temporarily ends hopes for agreed cessation of nuclear weapons tests—the one small step toward controlled disarmament which was, and still is, technically feasible. This proposal was first brought up in the *Bulletin* in articles by Colonel Leghorn and David R. Inglis.[4] It is usually supported by two arguments: One is that further radioactive contamination of the earth's surface should be prevented; the other, that the deadlock in disarmament negotiations could be broken by a test ban. This writer has been rather skeptical of the validity of either. The danger of test explosions to health and heredity (to which Professor H. J. Muller was the first to draw legitimate attention[5]) has since been proved to be too small to play a decisive role. The hope that cessation of weapons tests could break the deadlock in disarmament negotiations was, and remains, unfounded, as long as there is no manifest intention to change radically the whole approach to the problem of disarmament, because the cessation of tests was, and remains, the only disarmament step which could be effectively controlled without a far-reaching revision of national sovereignties. There were, however, two more arguments in favor of the cessation of tests presented by Colonel Leghorn. One was that an effective cessation of tests would prevent the United States, as well as the Soviet Union, from developing nuclear warheads for long-range missiles and thus stop the arms race short of the point of no return. The other was that cessation of tests

[3] C. W. Sherwin, "Securing Peace through Military Technology," and Warren Amster, "Design for Deterrence," *Bulletin of the Atomic Scientists,* XII (May, 1956,), 159–65.

[4] David R. Inglis, "National Security with the Arms Race Limited," *Bulletin of the Atomic Scientists,* XII (June, 1956), 196–201.

[5] H. J. Muller, "The Genetic Damage Produced by Radiation," *Bulletin of the Atomic Scientists,* XI (June, 1955), 210–12, 230.

would prevent new nations from ever acquiring atomic weapons of their own making.

It was the third reason that was mainly responsible for this writer's unenthusiastic endorsement of the test ban proposal. It has now lost its validity. According to Mr. Khrushchev, the Russians already have thermonuclear warheads for long-range missiles; and Dr. Edward Teller told the Senate Committee that we have them, too. The last argument—that of preventing additions to the number of "nuclear powers"—had merit and seemed to have made some impression on both the American and the Soviet political leadership. However, if one sees the future world development as progressive consolidation of the bipolar world system—a state of affairs in which nations, willingly or willy-nilly, group themselves around one of the two protagonists—then the danger lies not in minor nations' making (and using) their own atomic weapons but rather in their being supplied with such weapons, openly or surreptitiously, by the Soviet Union or the United States. If, on the other hand, one believes, with Walter Lippmann, in the return of the world to a multipolar system, with several nations or groups of nations independently pursuing their own policies, then one can ask: How long will the latecomers agree to their exclusion from the "atomic weapons club"? In the London disarmament negotiations, France has already refused to stay in this position of inferiority, unless the "charter members"—U.S., U.S.S.R., U.K.—are willing to promise the gradual dismantling of their own atomic arsenals.

The "fourth country" argument has not swayed American defenders of tests in the past; and it will not sway them now, when other arguments in favor of test cessation have become invalid and their own arguments for further tests have acquired enhanced conviction. This is because, after the perfection of nuclear missiles, the logical next stage in the arms race is an attempt to develop weapons capable of intercepting and destroying them. Nuclear weapons in the megaton range seem to provide the only glimmer of hope in this field. As long as peace is going to be balanced precariously on mutual deterrence, to ask for the cessation of tests means asking for renunciation of the hope that "deterrence by active defense" can be developed to supplement "deterrence by retaliation."

Thus, one finds no convincing reason to oppose tests needed to develop defensive nuclear anti-missile missiles. What we have the right—and the reason—to ask for is that testing be restricted to the minimum necessary for this purpose. An internationally agreed ceiling on the annual release of radioactivity into the atmosphere remains a worthwhile, and not impossible, aim.

That the proponents of continued testing now have the field does not mean that they were always right. They were motivated (it seems to this writer) by the feeling, which is overwhelmingly strong and widespread in America (or at least, it was before the Soviet satellites!), that any agreed halt of technological competition with the Soviet Union would be unfavorable for America, because our strength lies in superior technological and scientific potential as contrasted with the brute manpower of the East. Any voice to the contrary, arguing that America stands to lose—and not to win—from a continued race in technological arms, was truly a voice crying in the wilderness.[6] It is obvious now, in retrospect, that stopping bomb tests (as well as long-range rocket tests), when it was first proposed two years ago, would have been advantageous to the United States and disadvantageous to the Soviet Union. However, this is now water over the dam.

Probably, most scientists in the world have set their hearts on the cessation of weapon tests. The abandonment of tests has become in their eyes a symbolic act, signifying abhorrence of atomic war and the decision to seek peace in atomic disarmament. It has been endorsed by the Pope, by Nehru, by Adlai Stevenson, by the German nuclear scientists, by the Federation of American Scientists, and by the more than two thousand American scientists who have signed Linus Pauling's appeal. It would be a grave and sad decision to give up this hope. But scientists, more than anybody else, must remain rational, and the call for cessation of weapons tests has lost its rational justification—at least, for some time to come.

Some may say that these are thoughts of despair. If we cannot look forward to progress in disarmament, except perhaps in areas which will not affect the capacity for instantaneous mutual destruction; if we are to stop calling for the cessation of nuclear weapons tests, at least as long as the development of anti-missile weapons remains the most urgent part of the arms race; if mankind is condemned to live indefinitely on the edge of the precipice, into which not only the rashness of a dictator but even the foolishness of a sub-

[6] E. Rabinowitch, "A Last Chance," *Bulletin of the Atomic Scientists,* XII (June, 1956), 187. "There is a tendency in America to believe that to stop the development of advanced technological weapons, even on a truly reciprocal basis, would, on the balance, damage the U.S. and favor the Soviet Union. It has been so often stated that the military strength of America lies in its technological leadership, while that of Russia resides in its inexhaustible manpower, that this is accepted as permanent. . . . However, the situation is changing. Ever since 1945, atomic scientists have pointed out that *in the long run* the existence of atomic weapons will bring more advantages to the Soviet Union than to America. . . . The rapid advance of Soviet atomic technology and military aviation is about to make this long-range prediction come true."

ordinate can plunge it—what can we do but simply live unto the day and hope for a miracle? What good is a *Bulletin of the Atomic Scientists* if it can merely show the dangers and discount all remedies?

The answer is that looking the crisis of our time in its face and saying that the methods proposed so far to deal with it are unrealistic or insufficient is not desperation. Rather, it is like the frank statement of a doctor that the patient cannot be expected to recover by the continued application of household remedies—taking aspirins and going earlier to bed—when what he needs is a major operation or (perhaps a more appropriate simile) a radical change in his dangerous way of life. There *are* things which can be done in the present world situation to kindle the hope that, at the end of the dark and dangerous way, there is a future worth living for. They are not easy, but at least they are things which our own country, and every other country which wants to join in the effort, can do by itself, without waiting for the consent of others. They are not new developments in military art or new strategies in the cold war. They are political decisions, not in current policies but in fundamental political philosophy. What is needed is to break with the tradition of past history—which twelve years of atomic energy have made ancient history—and to start thinking about the world of the future and working for its birth.

If we realize, completely and irrevocably, that mankind, divided into territorial units, each a law unto itself, has no future other than to live with the ever present threat of mutual annihilation, that the only way out of this predicament is the development of a true world community, then our life in the shadow of death will acquire meaning and purpose, and every step in our day-by-day policies can be made a step toward the clear ultimate aim. The dreadful reality of a rigid system of mutual deterrence and the tensions, hatreds, and suspicions which this system engenders do not make a favorable climate for proclaiming our faith in a future world community and dedicating our political work to it. We see every day how this system stimulates nationalism and breeds distrust even between close friends. But we cannot wait for a different climate to meet the challenge, or the climate will never change except for the worse; and even if events not of our making (for example, internal transformation of the Soviet dictatorship) should create more favorable conditions, we will not be ready to respond to them.

If we struggle through to the birth of a new spirit within ourselves and acquire the sense of a new mission for our country—the mission of leadership in a long fight for the unification of mankind—problems of strategy, economics, and politics will appear in a new light. The arms race will be seen as something we cannot escape for the time being but something we have to carry on with a minimum of saber-rattling, restricting our weapons to the kind and number sufficient to deter aggression and making our capacity to act, and our decision to do so, as clear as possible, to avoid all danger of misunderstanding, even at the expense of secrecy now considered so important for our security. We will establish a system of command—if possible, a common command with our friends—and put people in charge of it who will offer the greatest possible guarantee against any rash action, whether at the top or farther down the chain of command. We will discard the idea that we have to be ahead in *every* kind of armament, since leadership in the arms race will not be the criterion by which we will strive to acquire friendship and respect. Even if keeping an adequate deterrent power may require much greater national effort than we are making now—as well it may—we will not permit leadership in the arms race to become the chief concern of our political thinking, an obsession, an aim in itself.

In our dealings with the peoples under Communist rule, we will support all steps which could make their involvement in the world community more likely. We will pursue the broadest human contacts and economic and cultural co-operation. We will resist the temptation to retaliate in kind for every Soviet chicanery. We will get out of the absurd situation in which the Soviet dictatorship can proclaim that we are more afraid of the exchange of people, goods, and ideas than they are. We will accept the certainty that every open channel will be used by them as an outlet for their ideology and influence; it is a price we will be willing to pay to demonstrate to the people under Soviet rule that we believe in freedom of ideas, that we stand for the integration and well-being of all humanity. We will end the discrepancy between what our left hand and our right hand are doing, which now effectively defeats our policies. We will not first initiate world co-operation on atomic power or earth exploration, of which President Eisenhower and Admiral Lewis Strauss are justly proud, and then enforce ridiculous "precautions" against contacts with Soviet science: endless visa delays, provision of "bodyguards" for visiting Soviet scientists, and so on.

We will seek and promote new world-wide ventures, along the lines of Atoms-for-Peace and the International Geophysical Year. We will make up our minds to consider scientific and technological progress anywhere in the world a "good thing." Despite sober realization of the danger with which this progress confronts us when it increases the military potential of our competitors, we will not be diverted from the conviction that, in the last reckoning, every discovery in science and every invention in technology, wherever made, can be of benefit to mankind as a whole. In the same way, we will make up our minds that economic progress is a good thing, wherever and under whatever economic system it occurs. Of course, a nation in good health, with a productive agriculture and an efficient industry, is a more dangerous enemy than a backward country on the verge of starvation. But we must accept, once and for all, as a decisive motivation, the belief that world community will be easier to achieve when hunger, poverty, and disease disappear everywhere.

In our relations with the uncommitted countries, we will balance the doubtful advantage of their military commitment to our side against the more real advantage of their peoples' seeing us as a nation committed to the cause of world peace and world prosperity. We will go to the limit of our financial and economic capacities to help their reconstruction and progress, treating them as we would parts of our own commonwealth. We will welcome all paths of this co-operation, direct or through the United Nations, whichever these nations may prefer.

We will keep in mind that our industrial and agricultural capacity exceeds our capacity for consumption. We could not eat all the grain, meat, butter, and eggs we can produce even if the fraction of our population which is still underfed were to get all the food it needs. We can produce many more cars than we really need; only by artificial stimulation of interest in new models, and new frills, can our market absorb the millions of new cars our industry can deliver year in and year out. However much we deny it, we can maintain full employment and full utilization of our industrial and agricultural plant only by producing for waste, whether in the form of consumer's waste or in the form of accumulated military hardware which rapidly becomes obsolete. The same effect on the national economy would be achieved by bringing hecatombs to the gods of full production every year, consigning to the flames or to the depths of the ocean a large fraction of our annual output of goods!

The Soviet leaders are as yet not in the same trouble. They still face a bottomless pit of unsatisfied consumer demand and pretend that they can go on increasing production indefinitely without saturating it. We hope they will soon reach this desirable state; but by then—or even earlier—they will become serious competitors for the allegiance of underdeveloped countries, since they will have no difficulty in putting excess productive capacity to work for the world at large. At present, we are far ahead of them in this capacity; but

Syria, Afghanistan, and other countries suggest the shape of things to come. Our ability to rally the world by serving its needs will be lost if we do not put it to full use soon.

If we accept leadership in the movement of mankind toward integration in a single community, we will offer to go as far as our friends and allies, particularly the people of Europe, will be willing to go in the creation of economic, political, and cultural unity: free exchange of goods, ideas, know-how, and men, and common economic and political institutions. We will make clear our willingness to consider sympathetically every advance they may want to initiate in this direction.

Of course, all this is enormously difficult; but in contrast to the difficulties of even patchwork political agreements with the Soviet Union, or of effective disarmament under the present divisive world system, these difficulties are not objective; they are not beyond our reach. They exist within us; they are subjective, psychological. These steps are things we can do, if we decide that we want to do them and if our leaders show the imagination to lead us in doing them.

Can it happen? Certainly not, if we stick to the hope that salvation can be achieved by mild palliatives, by increased armaments at home and endlessly continued diplomatic bargaining abroad. Without a realization that nothing short of regeneration of mankind can solve the crisis of our time, the strength needed for this new beginning will not come. A heavy responsibility lies on our political leaders. They have been elected to keep the good acquired in the past; will they realize that this cannot be done in our time without opening the door wide to the new? Will they tell the truth if they see it, believing in the capacity of a free people to rise to the challenge once it is revealed in its full, naked greatness?

In his forty-year jubilee speech, Mr. Khrushchev said: "The capitalist world, doomed by history, is no longer able to proclaim aims capable of rousing and inspiring millions of people." If all we in America stand for is capitalism—or, as we prefer to call it, the economic system of free enterprise—he may be right. The hope of mankind lies not in any system of production, capitalist or Communist; it lies, in the age of science, in the creation of a world community able to prevent the misuse of the powers of science for the self-destruction of mankind and to make possible its full utilization to satisfy the common spiritual and material interests of all nations. Here is an aim which those who are free from the dead hand of dogmatic economic determinism can make their own; and millions everywhere will rally to it.

Hail and Farewell

The recent change of the American administration has produced two statements which stand as signposts at the crossroads of American policy: the farewell speech of President Eisenhower, delivered over television on January 17, and the inaugural speech of President Kennedy, pronounced on January 20 from the icy porch of the Capitol. Both are brief, moving, and eloquent, but one looks to the past and the other toward the future. This is so not only because one speaker is looking back over eight years of his stewardship, while the other is looking forward to the beginning of his presidency, but because one speech is the expression of the noblest quintessence of American conservatism, while the other is an eloquent expression of the equally American spirit of pioneering innovation. The retiring President looks for inspiration to American history; the future President proclaims his welcome to the challenge of a new age, unknown to his forefathers.

President Eisenhower sees the task of American political leadership in the preservation, through the storms of our time, of the principles of political and economic freedom. In his belief, if America would only keep its own image unsoiled and stand steadfast in opposition to the Communist evil, the storm of revolutionary change, which now obscures the skies, would subside, and "peace with justice" would prevail. President Kennedy sees the world in violent transition toward an unpredictable future, and not in a passing convulsion. He believes that America must go out in the world and assume the leadership of the revolutionary forces that are rattling at the foundation of the old order. The difference between the two Presidents is the old contrast between isolationism and world involvement. President Eisenhower's administration has been, in practice, as interventionist as any of its predecessors; but intervention after intervention, it has considered each one as an emergency measure, while keeping the unspoken hope of America's returning sooner or later to the blissful state of minding its own business. President Kennedy's attitude stems, on the contrary, from the recognition of America's irrevocable entanglement in the affairs of the world and an acceptance of a permanent responsibility to mankind.

Toward the end of his speech, President Eisenhower gave to the

From *Bulletin of the Atomic Scientists,* March, 1961.

American people two pieces of advice. One of them was to watch the growing power of the military and the industrial forces allied with the military. America, he said, has had in the past neither a large standing military establishment nor a large armament industry. The changes in military technology now force her to have both. We must watch, he said, that these new forces do not acquire undue influence on our policies. This is a remarkable statement to come from a general. It reveals the kind of pressure to which the President has been exposed in the last eight years—pressure for higher defense expenditure and for a more belligerent foreign policy. It shows that the source of his steadfast resistance to these pressures has been not only his allegiance to orthodox fiscal principles but, above all, his faith in the traditional principles of American statesmanship, with its aversion to militarism in domestic as well as in foreign policies. Many Americans who voted for Eisenhower's adversary in two presidential elections will wholeheartedly welcome this advice, given by a military President to his civilian successor in the White House.

Yet this first piece of advice cannot be dissociated from the second one, which followed in the same speech. Americans should also watch, President Eisenhower said, for undue influence by scientists and technologists on United States policies. The American government has been forced, he said—almost with a tinge of regret—to exert its efforts for the advancement of science and technology in America, because this is what the world situation calls for. But Americans should see to it that the spirit of science and technology does not unduly affect our national political thinking and decision-making, which should remain in the hands of those imbued with the traditional, humanitarian attitude of American statesmanship. This, then, must have been the result of President Eisenhower's exposure to the advice of scientists. Although he often followed their advice (for example, in the matter of cessation of nuclear tests), he did not fully understand and trust them. The military, the armament makers, the technologists, the scientists—these constitute, in President Eisenhower's mellowed wisdom, disturbing alien forces which threaten to adulterate the political image of America.

In President Eisenhower's mind, private enterprise belongs, with political freedom and peaceful policy, to the American tradition. The West Point graduate has become imbued with the classical liberal philosophy of the nineteenth century, in which economic and political freedom formed an indissoluble unity. He has surrounded himself with businessmen, in his cabinet as well as in his private life; to him, the American business community, rather than the American liberal professions, is the carrier of the true American tradition of individual freedom and initiative, of religious humanitarianism, of serving the community by serving themselves.

Of all conservative ideals, American conservatism is, perhaps, the least selfish. It does not involve the perpetuation of the arbitrary rule of a family, or a class, or a single national group in a multinational state. Although it opposes the tendency toward economic equalization inherent in the welfare state, it favors the economic advancement of all, having found the fullest scope for the enrichment of the business class in the broad increase of the purchasing power of the people as a whole. It stands for a society which social mobility has made nearly classless. It has made America practically immune to the philosophies of class hatred and class struggle.

However, all conservatism, after its time in history, becomes a source of danger. British conservatism, with its stable class structure and its firm belief in the legitimacy of the British Crown's rule over the "lesser peoples" throughout the world, made England the top nation in the nineteenth century and saved it from floundering in the first half of the twentieth. But its time came, too, and Winston Churchill, who did not "become Prime Minister to preside over the liquidation of the British Empire," had to begin this liquidation himself and watch others complete it.

Whatever ups and downs the human society may endure, it does evolve continuously, paced by new technological achievements and new spiritual realizations. The time comes when every conservatism must yield to the flow of time or risk being engulfed in a too long delayed explosion. The British conservatism has yielded in the nick of time. The French conservatism is still struggling, trying to extricate itself in Algeria without capitulation. That America has not owned colonies outright, that it has given independence to the Philippines and made Puerto Rico a test case of the industrialization of a backward society, has postponed the crisis. The events in Cuba and the ferment in South America are, however, signs that economic domination, if not liquidated voluntarily in time, can become as much of a danger to a great power as direct colonial rule. Not too long ago, existence of political freedom and national independence in Europe and America served as a powerful stimulant to the quest of peoples in Asia and Africa for their freedom and independence. Similarly, in our time, the rapid industrialization of the Soviet Union and China creates in the underdeveloped nations a yearning for similar rapid progress toward economic power and economic independence. The existence of a powerful Communist camp assures every leader of a nationalist rebellion that whenever he gets into conflict with Western powers he will have a powerful Communist camp on his side. It is up to the Western nations to avoid such conflicts and useless for them to blame the Communist leadership for offering assistance to those who find us intractable or unhelpful.

The conservative answer to the present world situation is to sit tight, to support every opponent of radicalism in every corner of the world—be it Congo or Laos, Iran or Iraq—and hope for the best. This policy can succeed on many occasions, as it did in Guatemala; in other cases, it may at least prevent the worst, as it did in Lebanon; but in the long run, it is unlikely to work. Sending fire equipment or dispatching firemen to all the areas of the globe where the fire of revolution has broken out could be an adequate policy, if a reasonable expectation existed that the winds fanning these fires—the revolution of rising expectations in the underdeveloped areas and the conflict between the West and the East—would soon die down; but this expectation is mere wishful thinking. In our time, all nations, including America, can be cunctators only at their own peril.

It is a remarkable testimony in favor of freedom of thought and public discussion that realization of the present danger of conservatism is gaining in American public opinion, despite the relaxing influence of national prosperity. President Kennedy did not win the election because of his attempts to arouse the nation to this danger or because of his calls for national sacrifice; but the fact that his alarmist utterances did not prevent him from being elected is significant. Americans did not call on their government to throw out the ballast of conservatism and to set the sails of the ship of state for a perilous voyage into the unknown, but they are not scared by calls for such a radical new departure, for accepting risks and making sacrifices. They have become sufficiently aroused, as a result of open public discussion, to be receptive to such calls; and this gives a resolute administration sufficient backing to start moving.

President Kennedy's inaugural speech shows his awareness of the critical situation in which America and its allies of similar economic and political tradition find themselves. What, for the conservative mind, were merely bothersome disturbances caused by evil forces abroad in the world are, for him, welcome opportunities in the search for a better future for man. Science and technology are the first of these forces. President Eisenhower was wary of them; President Kennedy is not. These forces need to be brought to bear on the achievement of high goals at home and abroad, not reluctantly, as servants who need to be kept under tight control, but wholeheartedly, as efficient helpers in both the formation of national policies and in their implementation.

In the presidential campaign, both sides emphasized the defensive tasks of American leadership. Practically every campaign speech was dominated by discussion of the threat of Soviet aggression against the West or of Soviet takeover of the underdeveloped parts of the world. Not only the need for a greater armament effort, but also the need for an increased and more sustained effort in assistance to underdeveloped nations, was motivated by the need to improve our defensive posture. Even our efforts in science, particularly in space science, were justified by the need to catch up with, or stay ahead of, the Russians. This looked like a bad omen for the coming administration. It seemed to presage a failure to break away from the defensive spirit that has dominated American policy in the last decade. Conservative spirit begets a defensive psychology; its natural inclination is to leave the initiative to others and merely hold steadfastly to its own. Unwillingly, it dispatches firemen to extinguish fires, but it does not send builders to build fireproof buildings. It may appear paradoxical that an administration elected on the slogan, "Enough of containment; let's roll the Soviet tide back," became in practice dedicated to a purely defensive policy; but this could have been foreseen because an inherently conservative administration will naturally choose the policy of lesser exertion and lesser risk. More vigorous defense, bigger fire engines, and huskier firemen, which seemed to be what Kennedy promised in the campaign, would not have produced a real break with the conservative past; it would merely have made the future more ominous. Kennedy's statements on Cuba in the election campaign were not reassuring.

The more remarkable is his inaugural speech. It places all emphasis not on the strengthening of our reaction to Soviet threats but on constructive policies initiated by ourselves. It calls Americans to work and, if need be, to sacrifice to give America leadership, not in the fight to keep down the national revolutions in the new nations, but in the effort to help these nations to achieve true independence and economic viability. Instead of presenting our help to technically backward nations as a weapon to keep the Soviet influence out of them, he boldly suggests that the Soviet Union join us as a full partner in the technical assistance program.

In many articles in the *Bulletin of the Atomic Scientists,* I have argued that the only hope of mankind to avoid, in the long run, a military showdown between the two powerful camps into which the world is now divided is to break with the age-old tradition of national policies dictated by national interests and to subordinate the interests of separate nations to the common interests of mankind. Such common interests have always dominated pure science and, to a large extent, many of its applications, such as medicine. Now, with the increased importance of science and technology in human affairs and the boundless prospects offered by the systematic application of science to the satisfaction of human needs, the time has come to recognize that the pursuit of common interests is more important for every section of mankind than the traditional advancement of its own interests. It appeared to me that assistance to underdeveloped nations is the one field in which the interests of all powers should coincide. Educational and technical rehabilitation of these nations is more important than hopes to attach these countries to one or the other of the contending camps. The alternative course, that of competition for the allegiance of the underdeveloped nations and their consequent division into two camps, is fraught with the danger of fostering uncounted local clashes, each of which could become the spark igniting a general war.

The idea of co-operative assistance to underdeveloped nations, with equal participation of all technologically advanced nations, has found its expression in discussions between Western and Eastern scientists at the several Pugwash meetings. It was unanimously endorsed in the "Vienna Declaration," September, 1958,[1] and reaffirmed in Moscow, December, 1960. The Vienna Declaration stated:

> We believe that the traditions of mutual understanding and of international co-operation, which have long existed in fundamental science, can and should be extended to many fields of technology. . . .
>
> The extremely low level of living in the industrially underdeveloped countries of the world is and will remain a source of international tension. We see an urgent need to forward studies and programs for the effective industrialization of these countries. This would help reduce the sources of conflict between the highly industrialized powers. Such studies would offer fruitful scope for co-operative efforts between scientists of all nations.
>
> We believe that through such common effort the coexistence between nations of different social and economic structures can become not merely peaceful and competitive but to an increasing degree co-operative and therefore more stable.

In pursuance of this aim, the Continuing Committee of the Pugwash conferences has been working for the last two years toward an international meeting of scientists and technologists, to discuss the possibility of world-wide co-operation in the advancement of science and technology in the new nations.

The International Atomic Energy Agency, established by the United Nations after President Eisenhower's "Atoms-for-Peace"

[1] Included as an addendum to Part III of this book.

speech, and the United Nations Special Fund have not been spectacular successes, but they are encouraging signs of the progress of international co-operation in the assistance to underdeveloped nations on the governmental level. Occasionally, proposals to seek Russian participation in international assistance activities were brought up by world political leaders, particularly by General de Gaulle. President Eisenhower has suggested such co-operation, at least in the field of medicine. Adlai Stevenson made a wider proposal of the same kind in a Roosevelt Day speech a few years ago. However, these half-hearted utterances were lost among loud arguments emphasizing distrust of the Soviet leadership, and the need for military and technical assistance to offset the threat of Soviet expansionism, and often opposing our help to nations not willing to join the anti-Soviet military camp. The practical demonstration of the possibility of constructive co-operation with the Soviet Union in applied science—the International Geophysical Year—has been forgotten in the bitterness of the Soviet-American controversy that followed the U-2 crash in the Urals on May 1, 1960.

It is, therefore, a matter of great significance that President Kennedy in his inaugural speech committed himself clearly and unequivocally to the pursuit of the common interests of mankind as a prime objective of United States policy and invited the Soviet Union to join in these efforts. Proclamations are not policies; any attempt to convert Kennedy's appeal into practical policy will encounter obstacles from both sides. Some of these obstacles are psychological, the result of ideological prejudices which make people blind to the existence of the common interests of the Communist and the libertarian societies. It is widely believed that what is good for one is bad for the other, that any offer of common effort from the West must be a deception intended to subvert the East, and vice versa. In fact, both sides are so deeply committed to the pursuit of their own "righteous" purposes, and opposition to the "evil" aims of the other side, that some kind of a temporary practical schizophrenia will be necessary for them to act creatively together. Islands of peace will have to be established in the tumultuous ocean of conflict. The success of the IGY is proof that such double standards in international relations are not impossible. The problem is to make the areas in which creative co-operation is possible loom larger in the minds of people than the areas of continued conflict.

The second set of difficulties is real and not psychological. There is no region in the world in which our practical interests and policies are not in conflict with those of the Soviet government. In Berlin, in Eastern Europe, in southeast Asia, in Africa, in Latin America, this antagonism expresses itself in a variety of ways, and each variation threatens mankind with a possible outbreak of open warfare. How can we combine the pursuit of our interests, which are in such conflict with those of the Soviet Union, with a co-operative effort toward the rehabilitation of these regions? We have a live example of this difficulty in the Congo, where the efforts of pacification and rehabilitation initiated by the United Nations, with the consent of both camps, is half-paralyzed by the contest between the Soviet-oriented Lumumba followers and the Western-oriented Kasavubu-Tshombe alliance.

Kennedy's inaugural speech falls short of facing this central problem of our time. It says nothing about the need for stabilization of the world situation, which is a prerequisite for systematic, successful co-operation of the East and the West. It is also a prerequisite for successful disarmament, another subject on which the inaugural speech was silent. For several years I have argued that political stabilization (which, at the present time, is possible only on the basis of recognition of all existing political realities in the world) is a *sine qua non* of a realistic disarmament program.

Constructive co-operation of East and West in pursuit of common ends of humanity, disarmament, and stabilization of the political situation in the world are a triad of political aims which cannot be pursued separately. If anything, political stabilization should have the first order of priority. I know that on this last point I am in conflict with the opinions of most of my colleagues, both in America and in the Soviet Union. At the recent scientists' conference in Moscow, Mr. Noel-Baker movingly expressed his belief that the arms race is the primary cause of all wars and that disarmament would permit easy and peaceful solutions of all political conflicts. The whole discussion of the disarmament problem in this conference (as well as in official negotiations) seems to be predicated on the belief that great progress in the disarmament area can be achieved without substantial progress in political stabilization. I wish I could believe it, but I can't. Disarmament seems to me possible only among nations which have found that their common interests are strong enough to dampen the pursuit of their contradictory, separate interests, both in their general attitude and in their specific policies in various areas of conflict.

Since I have repeatedly discussed this matter before, I will not pursue it here, except to say that the dilemma cannot be avoided. The Dulles policy of preventing political stabilization of the world by refusing to recognize the *de facto* frontiers in eastern Europe, by encouraging the belief that we will support anti-Communist revolutions (despite our signal failure to do so in Hungary), and by helping the opposition to neutralist regimes in southeast Asia, and so on, could be justified if there were reasonable expectation of its success in the foreseeable future. In the absence of such expectation, our refusal to stabilize the existing situation merely paralyzes all programs of building a peaceful world. President Eisenhower liked to speak about "peace with justice," meaning peace which would give every nation the right of freely choosing its political and economic regime. This aim of "self-determination of nations," first proclaimed by Woodrow Wilson, remains a noble ideal. But Woodrow Wilson proclaimed it at the time when the United States was fighting, and winning, a war and when it had considerable (although ultimately unused) power to impose its ideas on the settlement ending that war. Proclaiming the same ideal now, when we do not fight and do not intend to fight for its realization, can be only a source of self-righteous satisfaction.

It is reasonable to suppose that the new administration is not unaware of the logic of these considerations. The sudden anxiety of Chancellor Adenauer, the main protagonist of the policy of "non-recognition" in Europe, to come to terms with Poland; the rumors preparing the world for a Western recognition of the Oder-Neisse border between Germany and Poland—all these may be symptoms of the direction in which the thinking in Washington is pointing. The stabilization of the situation in Berlin (as well as in Taiwan) is only possible in the framework of general abandonment of the non-recognition policy. Certainly, opposition to this change of policy, at home and abroad, is bound to be widespread and violent. It is to be hoped that the new administration will have the clarity of purpose and the practical political wisdom to achieve the turn with what has become known as "deliberate speed."

15

New Year's Thoughts—1962

Christian Morgenstern, author of *Galgenlieder* ("Songs from the Gallows") defined the working of a dogmatic mind in the famous lines: "Weil, so schliesst er messerscharf, nicht sein kann, was nicht sein darf!" (For, so sharply argues he, it cannot be what should not be!). A resolution of the American Air Force Association of September, 1961, asserts that the United States, with its superior power (and the justice of the cause for which they—and we—believe it stands), should not be incapable of destroying the weaker (and wicked!) Communist power! Therefore, this incapacity cannot be; America's apparent helplessness in its struggle with the U.S.S.R. must be the result of our lack of will to mobilize and use American power; given this will, the United States must be able to win the cold war (and, if need be, a hot war, too).

Another spokesman for the return of "realism" in foreign policy, Mr. John K. Jessup, in an article displayed in three issues of *Life* magazine, gives the following prescription for winning the cold war:

> *One,* build a bigger U.S. military capacity . . . and formulate a new doctrine for the use of force. *Two,* create new institutions to make possible unified cold war action by the Atlantic community. *Three,* organize concerted propaganda policy, especially within the U.N. *Four,* establish concerted allied trade policy for economic warfare. *Five,* institute a sounder policy for foreign aid, both for political and economic aims.

Mr. Jessup does not indicate why and how these policies will lead to a victory in the cold war; but since, in his mind, continued existence of communism is something which *should* not be, the readers are asked to believe that communism will somehow disappear from the earth if Mr. Jessup's political advice is taken.

Of course, Communists themselves display an even blander reliance on eventual disappearance of things which, according to their book, should not be. Some time ago, Trofim Lysenko sold his quack theories to the Communist leadership with the argument: "Communism needs the capacity to transform quickly the species of plants and animals; genetics says that this cannot be done; *ergo,* genetics must be wrong." The political planning of the Communist leadership is based on a similar proposition that what they call "capitalism" should not and therefore cannot survive, all evidence of the

From *Bulletin of the Atomic Scientists,* January, 1962.

resurgent viability of non-Communist economic systems notwithstanding.

The "historical realists" who extrapolate from the past history of mankind that the conflict between the Communist and the non-Communist camps can and will be decided by a test of force, and that the side that should win it will win it, simply refuse to admit the facts created by the development of science and technology, which make the fighting of a war irrational and suicidal for both sides—the stronger as well as the weaker, the righteous as well as the wicked. These rigid minds maintain that a stalemate which should not exist cannot exist, and they call for the issue to be joined and the enemy to be utterly destroyed.

Two years ago, the clock on the cover of the *Bulletin of the Atomic Scientists* was moved back a few minutes in recognition of the fact that public opinion and governments in all parts of the world had begun to show some understanding of the facts of life in the atomic age and of the impossibility of continuing into this age the traditional power politics of the past. Listening to the rising chorus of American advocates of stepping up the power contest, of "winning the cold war," and of "destroying communism," one begins to doubt whether this slowly acquired understanding of the facts of the scientific age is not being submerged by the atavistic urge to end a prolonged and bothersome conflict by bodily destroying the opponent. Children from primitive tribes, after education in advanced societies, have been reported to be rapidly engulfed by the traditional ways of their society once they return home from school. Are we witnessing a similar return to primitive behavior of our political leadership and public opinion after their exposure, for fifteen years, to schooling in the facts of nuclear warfare?

One thing is certain. If reason (or fear) prevails this time, and the conflict over Berlin is somehow patched up, a new and long-lasting effort of "education to reality" will be needed to avoid the repetition of similar crises in the future. This education will require more than retelling again and again the physical and biological consequences of the use of nuclear weapons. It will have to aim at a re-evaluation of national values—a shift of emphasis from national aims which can be obtained only at the risk of nuclear war to aims which lead away from the power contest and war threats.

The Soviet Union, in particular, will have to learn a lesson. It will have to realize that any attempt to change, by force or threat of force, the political and military status quo anywhere in the world, and particularly in Europe (even if this change is attempted, as in Berlin, in the name of stabilization of this status quo), is fraught with the danger of nuclear war. Such attempts are likely to sweep away the slow and painful recognition, in the West, of the need for accommodation and co-operation between the East and the West which has grown during the "armistice" that has prevailed between the Korean War and the Berlin conflict. Even if additional stabilization of the status of Eastern Germany is secured by the Soviet Union in the settlement of the Berlin crisis, will this gain be worth the exacerbation of Western public opinion and the strengthening of militant anticommunism in the world that this Soviet excursion has produced? The time allotted mankind to evolve from precarious armistice to stable peace is limited, and much of it has been irrevocably and unnecessarily lost by the arbitrarily provoked conflict over Berlin. Perhaps we can hope that the apparent victory of pragmatism over dogma in Soviet society will make it unnecessary for the Soviet government to indulge again in aggressive gambits to prove its Communist "reliability."

What is the truth about our time? How will it be described by historians a thousand years hence, if there are historians? Many Americans imagine that these historians will describe our era as that of a decisive conflict between the forces of political and economic freedom and those of totalitarian dictatorship, represented alternatively by Hitler's "Thousand-Year Reich" and the Communist juggernauts of Moscow and Peiping. The Communists, on the other hand, anticipate that future historians will look on our time as the era in which a new economic order, based on state ownership of the tools of production and therefore capable of effective long-range, centralized planning, was engaged in a decisive battle against the old order of economic anarchy and exploitation of the majority of workers by the minority of tool owners. Political freedom, on the one hand, and public ownership of the tools of production, on the other hand, appear to these two groups of close-range observers as the central phenomena of our time.

But overshadowing these two alternatives, however important they may be, will be, in the eyes of future historians, the scientific revolution. In gestation for two or three centuries, it has erupted in full force since the Second World War. It is only because of this revolution that the conflict between the Communist and the non-Communist worlds has assumed its apocalyptic aspects; it is because of this revolution that the contemporary awakening of the economically underdeveloped parts of the world has become fundamentally different from the emergence of new powerful races and nations in the past: the Meiji reformation in Japan, the sudden upsurges of Macedonian or Mongolian power, the waxing and waning of the Persian, Aztec, and many other civilizations. The scientific revolution replaced such localized and isolated phenomena with a world-wide "revolution of rising expectations," sustained by a reasonable belief in the possibility of a satisfactory life for all peoples. This one consequence of the scientific breakthrough alone will loom larger in the eyes of future historians than the conflicts between two political and economic systems which now divide Western civilization.

To our descendants, our era will appear above all as the era when man's life expectation, whether he lives in one of the stable libertarian societies of the West, in the rigidly disciplined Communist society of the East, or in one of the emerging countries of Africa or Asia, has increased two or three times in the span of a few generations—an era when the amount of energy available for the average individual to satisfy his wants and his desires has grown tenfold, if not a hundredfold. In this era, man has penetrated the atom and released the awesome energies of its nucleus; he has first left his planet to surge into space. This revolution in human existence unfolds before our eyes, not as part of the conflict between the two camps into which the Western world is divided, but, in essence, through their common efforts. Our era may appear, to a myopic mind aware only of the bitter conflict raging between the ideas extrapolated from mankind's obsolescent experience, as an era of alienation in which mankind has become separated as never before into two irreconcilable camps; but to future generations it will appear as an era of beginning world-wide co-operation of mankind. Never before have so many outstanding men of all nations devoted themselves to parallel or concerted activities based on commonly acknowledged truths and dedicated to the common aim of better understanding and more effective mastery of nature.

This is said not to dismiss lightly the ideas of political freedom, national independence, and social justice, which now appear to be of paramount importance to leading minds in all countries. These are great ideas; their attempted realization has been among the great achievements of history; they are worth the dedication of individuals and nations. But the scope of the scientific revolution of our time is so immense, and so pregnant with still greater future potentialities, that it is transforming the very bases of human existence—the background against which these political and economic ideas must be evaluated. This revolution puts an end to forces which have created, in the past, the bitterest conflict between nations and classes. It is destroying the justification of aggressions and conquests, which have been in the past the normal means by which nations could increase their power and prosperity; it destroys the rationale of exploitation of man by man as the only means of maintaining a high standard of life for certain small social groups. It opens the path to a healthy and economically secure existence for all men everywhere.

At the same time, the scientific revolution creates new challenges of which men are largely unaware. The challenge of overpopulation; the challenge of human leisure (which is recognized as acute in the United States and considered imminent in the Soviet Union); the problem of the increasing age of the population; the problem of work motivation in industrial civilization, in which the individual is widely separated from the product of his work; the problem of deterioration of the national genetic endowment through elimination of natural selection; the problem of growing urbanization and the disappearance of the rural way of life—all these challenges face all leading nations, irrespective of their economic or political systems. Those who have no will or capacity to look beyond a narrow theoretical framework based on obsolescent experience can be blandly convinced that all these problems of the scientific age could be routinely solved if only their theories would prevail; thus Soviet sociologists believe that the problem of overpopulation will never exist for Communist nations. But scientists all over the world should know—and many of them do know—differently. They know that the basic challenge facing mankind in our time is not the choice of a certain economic or political system but full utilization of all potentialities of the scientific revolution for the benefit of man and avoidance of their utilization for the mutual destruction of mankind.

Scientists are aware that mastering the challenges of the scientific age will require a new development of ethical and moral forces in human society beyond that needed for the establishment of political and social justice—the challenge which has preoccupied human minds in recent history. Contemplating the wide sweep of the scientific revolution in our time, and the fascinating prospects and difficult challenges which mankind faces because of this revolution, should not and will not destroy man's loyalty to the principles in which he believes, such as the principles of individual freedom and responsibility in Western civilization; but it should widen his perspective and prevent him from seeing in the momentary ups and downs of the now raging conflicts the essential content of human existence. Some people—like Hitler in his last days—can convince themselves that the annihilation of their nation, if not of all mankind, is preferable to the defeat of their political ideals or power aspirations. They consider those who—like Bertrand Russell—refuse to risk the existence of mankind to prevent a setback of the

West in its conflict with communism as cowards and traitors. These people fail to see that the conflict between the two political and economic systems of our time is but one aspect of a much greater transformation in human existence. They do not see that the future of their societies and their ideals depends, above all, on man's capacity to master the forces unleashed by the scientific revolution for the common benefit and to avoid self-destruction by their misuse; that if man succeeds in mastering this challenge the ideals of freedom, on the one hand, and of economic justice, on the other, will survive and grow, whatever the next phase of the conflict between the Communist and libertarian societies may bring. The ultimate fate of libertarian ideals in human society depends, not on the vagaries of the present ideological and political contest, but on the capacity to combine political freedom and the rights of the individual with full acceptance of the transformation of human society that the scientific revolution is bringing about, with the emergence of forces and the growth of concepts that require for their application organized co-operation of large masses of men rather than spontaneous efforts of individuals, and with the rapidly growing capacity of man to influence other men's attitudes and thoughts. Individual freedom and freedom of thought will have to find defense and protection against the dangers which these freedoms have created and which may be more crucial for their survival than the Soviet political and economic offensive.

The new year 1962 finds mankind in the midst of an upheaval of much broader scope than the mere conflict between the East and the West over the domination of certain areas in Europe and Asia, in the face of a challenge much greater than that of maintaining free communications with West Berlin or protecting South Vietnam. The truly great challenge of our time is a challenge common to all mankind, to libertarian as well as to totalitarian societies. It has been created, and must be answered, by the common efforts of all men.

The need to see the political and military conflict of the present time in the perspective of the great sweep of the scientific revolution, of course, applies to Communist society as much as if not more than, to the libertarian society of the West. Each side in the immediate conflict has its reasons for trying to solve it on conditions most favorable to it; but neither will perish by making concessions. The survival of both systems will depend on avoidance of war; their success, on their performance in satisfying men's urges for peace and for a full, free, and creative life.

PART III
TALKING WITH WORLD SCIENTISTS

Introduction to Part III

Scientists of the world form a single community, speaking a common language, following a common code of procedure. This community embraces scientists from the Communist East as well as those from the libertarian West.

Soviet science is a continuation of Russian science and as such a major branch on the tree of modern science. Its traditions are common with those of science in Western countries. This common origin could only temporarily be obscured by an ideological indoctrination and attempts to subordinate science to an outside system of political and economic values. This does not imply that the Soviet scientific community is not, in the great majority, loyal to the society to which it belongs or that many scientists do not see in communism an economic system permitting the fullest possible utilization of science for the satisfaction of human needs, unhampered by the contradictory influences at work in a system of private enterprise. The history of the precipitous growth of Soviet science, outpacing the rapid development of Soviet industry (not to speak of the halting advance of agriculture), supports this belief. In the eyes of many Russian scientists, communism offers the most promising way to transform the harsh reality of Soviet existence today into a bright future. N. N. Semenov, the Russian Nobel prize winner in chemistry and chairman of the million-member Soviet Society for the Dissemination of Science, sees a time in the future when every man and woman in the Soviet Union will be devoting part of their time to science.

It cannot be the object of Western scientists to question this loyalty or discourage these hopes, no more than we would let Soviet scientists question our belief in the importance of personal liberty for the spiritual and material advancement of mankind or subvert our dedication to political freedom, as aspired to and partly achieved in the West. The ground on which both can meet is common belief in the great role of science and technology in the progress toward a better life in all societies, irrespective of the prevailing social and political systems, and common interest in preventing science from being used for the mutual destruction of nations.

Four articles in this part of the book are attempts to discuss world problems in this light with scientists from many countries, including the Soviet Union. They are based on papers presented at several international "Conferences on Science and World Affairs" (COSWA, or Pugwash, conferences). Collections of all papers presented at these conferences were distributed to all participants and also made available to the participants' governments. These conferences (of which ten have been held to date) began in July, 1957, in response to an appeal drafted by Bertrand Russell for scientists of the world to "assemble and discuss" problems created by the release of atomic energy and its application to warfare. This appeal was signed by Albert Einstein—his last action before he died—and several other outstanding scientists. The first conference was held in Pugwash, Nova Scotia, on the estate of the Cleveland industrialist, Cyrus Eaton; this is how the conferences acquired their name.

Two years earlier a group of four leading Soviet scientists had appeared at a conference on the same subject, assembled in London by the British group of Parliamentarians for World Government; but afterward, attempts to continue communications with them proved futile. It was, therefore, an encouraging sign when, after the first Pugwash conference, its conclusions were endorsed by the Soviet Academy of Science and given wide dissemination among Russian scientists.

The second conference was held at Lac Beaufort, Quebec, in June, 1958, and was specifically concerned with disarmament. It was held in private, and no public statement was issued at its end. To the participants, it showed that scientists, coming from all parts of the world, could discuss various aspects of disarmament with objectivity and mutual comprehension; it justified the hope that continued discussions of this and other problems of common interest might help in the evolution of public thought and official policy on both sides toward greater understanding of the challenges of the scientific revolution.

The Lac Beaufort conference became the prototype for several subsequent ones: the fifth Pugwash conference, in Baden, Austria, in September, 1959; the eighth, in Stowe, Vermont, in September, 1961; and the ninth, in Cambridge, England, in August, 1962. A relatively small number of scientists took part in these conferences, selection being based, at least in the West, primarily on first-hand familiarity with the problems of the arms race and disarmament, which had been acquired in their work or from advisory activity to their governments. The aim was for these men to discuss the problems frankly, attempting to reach mutual understanding rather than common consensus.

Other conferences—the third conference, in Kitzbühel, Austria, in September, 1958; the sixth, in Moscow in December, 1960; the seventh, in Stowe, Vermont, in September, 1961; and the tenth in London in September, 1962—had a broader scope and wider attendance, and public statements were issued at their conclusions. The "Vienna Declaration" of September, 1958, issued at the end of the Kitzbühel conference, is reprinted as an addendum to this part of the book. It was drafted by the author, but incorporated amendments by many participants, until it became acceptable to practically all of them—conservatives, liberals, communists, pacifists, and arms-developers. It thus summarizes the common beliefs which a great majority of scientists hold (and to which they were willing to commit themselves publicly!).

The seventh conference, in Stowe, produced a report on "International Co-operation in Pure and Applied Science." Scientists from the West and East participated with equal exhilaration in the development of this program. One section of it suggested international co-operation in space research and exploration, which has since become a subject of official negotiations; another, the creation of international laboratories equipped with uniquely ambitious instruments—particle accelerators, reactors, computers—for physical and biological research; a third, a world-wide survey of the oceans, the earth's crust, and the atmosphere, and an inventory of the natural resources of the globe; a fourth, co-operation in technical assistance to underdeveloped countries; a fifth, the pooling of abstracting, information-retrieving, and reviewing activities in science.

Soviet scientists come to the COSWA conferences with less freedom than their Western counterparts; but one is encouraged to see this freedom grow with time. In contrast to Western scientists, Soviet colleagues do not criticize policies of their government or disagree with views which have been formally endorsed by the political leadership; but they discuss freely many aspects of arms control and world security, the responsibility of scientists in our time, and international co-operation in science and technology—and they do so with considerable enthusiasm.

The COSWA conferences are based on the belief that certain common attitudes of scientists—which include awareness of and respect for facts, commitment to unprejudiced thinking, and readiness for experimentation—constitute a not negligible asset in the present crisis. About the strength of this force, one should make no exaggerated assumptions. However, the hope that it may have a significant influence on the future behavior of nations is not implausible, because the facts of our time drive the governments, however unwillingly, toward the same conclusions which scientists derive from their particular acquaintance with the realities of the scientific revolution.

Societies are beginning to recognize, however confusedly, the incompatibility between the traditional attitudes which require the absolute loyalty of citizens to states pursuing their separate national or ideological interests and the immense power which science is giving to these societies. Scientists everywhere have one common responsibility: to educate their nations to the facts and prospects of the scientific revolution and to smooth in this way the path toward understanding among nations, now separated by deep moats of conflicting ideologies and nationalisms. This is difficult because it calls for a change of inbred habits and stubborn convictions. It calls for the abandonment of attitudes which have emerged from the adaptation of societies to the challenges of past history. And yet, such a reorientation has begun, both in the West and the East. The race between the growth of a new world-wide community, the only kind of society viable in the era of science, and the drift toward a scientific war, destroying all civilization—if not mankind as a whole—is on, and its outcome may remain uncertain for a long time to come. It is the task of scientists to increase the chance of mankind's winning its race against death, however small this chance may appear at the present time and however weak their direct influence on political developments may be.

International Co-operation of Scientists

Slowly, international co-operation in science, disrupted by the war, is being resumed. Conferences are being held, partly on the initiative of long-established international organizations and partly under the sponsorship of newly created United Nations bodies such as UNESCO. Every scientist is wholeheartedly in favor of full co-operation with his fellow scientists abroad. For a long time after the age of the Enlightenment, when the international scientific community was first established, this community continued to function even in times of war. Governments warred on each other, but academies remained at peace. But the growth of nationalism, together with the development of modern technology, has gradually transformed the limited wars of dynasties into total wars of nations. Now, after two world wars, the development of scientific methods of warfare has chained science tighter than ever to the national state. As a consequence, high walls of secrecy are being erected, not only between former enemies, but even between recent allies.

Every international scientific conference is a welcome event. Not only do these reunions bring together old friends who remember the brotherhood of science of before 1933—perhaps even before 1914; what is more important, they bring the tradition of science as a unifying force in humanity to the conscience of a new, war-grown generation of scientists. Only by a return to the old universal tradition can science escape becoming a handmaiden of national technology—war technology, above all. Such a utilitarian nationalistic degeneration would not only deprive science of its moral value as a disinterested search for truth but also deter great minds from entering it and thus write an end to the age of scientific discovery.

However much we may welcome the restoration of professional co-operation between nations, it is not enough. The atomic bomb, germ warfare, the world-circling stratosphere rockets, will not wait for gradual restoration of international trust and co-operation. Unless a new departure in international relationships is made soon, all the well-meaning attempts to resume international scientific life will come to naught. The apprehension of a catastrophic development has caused large groups of American scientists to forget their past aloofness in politics and to organize for a fight to prevent science

From *Bulletin of the Atomic Scientists,* September, 1946.

from becoming an executioner of mankind. Within a year, this fight has brought two successes: the passage of a domestic law of atomic energy, conceived in the spirit of peace, and the official adoption by the United States of a program of international control and co-operative development of atomic energy. The American scientists are overwhelmingly behind this program, not because as loyal citizens they support the policy of their government, but because they are convinced that international control offers the only alternative to an armaments race.

American scientists may, by themselves, achieve further useful results in the enlightenment of the American people. They may influence American international policy in the field of atomic energy by stressing the primary importance of this subject in world affairs and by supporting a patient, sustained effort to bring about understanding among nations. But what is most needed now is a broadening and internationalization of the fight for a rational solution of atomic energy problems. We need an international organization of scientists dedicated to this fight. It must be formed in a truly scientific spirit of search for objective truth and devoted to service to humanity, transcending national, party, or class affiliations; and it can only achieve its purpose if it brings together the most representative scientists of the world.

During the past year, when plans for international control were discussed among American scientists, many anticipated that these plans might appear in a different light to Americans and to sincere friends of America abroad. This was to be expected particularly in the proposed step-by-step procedure of disarmament and release of information. It was felt that, in such a case, the initiative for formulating alternative proposals might best be left to representatives of other nations. Some of these divergent views are now coming to light. We believe that a reasonable program for their solution can be worked out in a frank discussion between qualified representatives of world scientists. The realization of this program must be the common goal.

We need international organizations of scientists to attain various worthy objectives: advancement of scientific knowledge, furthering of the welfare of scientific workers, study of social problems of world-wide importance, and dissemination of information on scientific developments of major importance to mankind. But above all this, we need an international community of scientists to develop a positive program of international control of atomic and other weapons of mass destruction and to lead national groups in a concerted effort for the realization of the program.

Russian and Soviet Science

The recent outstanding successes of Soviet technology and science can be baffling only to those who forget that Russia has had a long and honorable tradition of scientific achievement dating back to the early eighteenth century. The most famous of early Russian scientists was Mikhail Lomonosov (1711–65), who was instrumental in making the Russian Academy of Sciences, founded by Peter the Great in 1724, a great and justly renowned scientific institution. He was the precursor of many other men whose contributions in the succeeding two centuries placed Russia prominently on the scientific map of Europe. Lobachevsky and Chebyshev in mathematics, Mendeleyev in chemistry, Lebedev in physics, Ivanovsky, discoverer of the first virus, in plant physiology, Dokuchaev and Pryanishnikov, who originated soil science, Zhukovsky, who pioneered in aerodynamic research, Pavlov and Mechnikov, whose contributions in biology and medicine were rewarded with the Nobel Prize—all these names surely testify to the fact that Russia could match the advanced countries of western Europe in its achievements in many fields of scientific endeavor.

To be sure, after the mid-nineteenth century, the backwardness of Russia's ruling classes hindered and delayed progress in academic and applied science. The best scientific minds had to work in small, poorly equipped laboratories with little official encouragement. Nor did the growing conflicts between the reactionary government and the liberal intelligentsia fail to touch the scientists. Many prominent ones, among them Mechnikov, Sophia Kovalevskaia, and Bakh, went abroad. Even the conservative-minded Mendeleyev had his share of troubles with the government over academic freedom and the autonomy of the universities. Nevertheless, when the revolution came, the ground was already well prepared for further scientific advance. A tradition of high-grade, forward-looking research was firmly established at the leading universities—St. Petersburg, Moscow, Kharkov, Kiev, and Kazan—as well as at various technological institutes and specialized military academies. The overthrow of the tsarist regime encouraged widespread hopes among scholars and intellectuals for the advent of a new climate of freedom favorable to academic and scientific progress.

From *Problems of Communism,* VII (March-April, 1958).

These hopes were soon disappointed. After seizing power from the short-lived democratic provisional government, the Bolsheviks established a dictatorship and proceeded to change everything by force. Laboratories stood empty, without gas, heat, or electric power. Students went off to fight in the civil war campaigns. Most professors, seeing their cherished dreams of freedom and representative government trampled upon, adopted an attitude of non-co-operation toward the Communist regime. Many scholars fled to the regions still occupied by the Whites, later escaping abroad.

Bolshevik policy, however, soon began to discriminate between the natural sciences, on the one hand, and the social sciences and humanities, on the other. At an early stage the regime proclaimed its intention of fostering the advance of science and staking the future of Communist society upon the power of science to transform the economy. It implemented this intention by offering material inducements and public recognition to scientists, and co-operation by the latter was rendered easier by the fact that it did not entail the same moral capitulation that was required of historians, economists, and philosophers. The result was to place scientific talent back in charge of teaching and research.

For the first fifteen years of the Communist regime, science continued to develop under relatively favorable conditions, an oasis of intellectual freedom in the desert of compulsory conformism. The ruling party largely refrained from restricting the choice of subjects of scientific study and the objective interpretation of results. This freedom, coupled with economic encouragement and social recognition, led to a rapid expansion of scientific education. The pent-up desire for knowledge released by the social revolution flooded the classrooms and research laboratories; eager but poorly prepared young men and women aspired, as the saying went, to bite into the granite of science.

There were, however, numerous difficulties. Apart from the deficiencies in preliminary educational preparation, there was a shortage of trained scientists to serve as teachers. Many had to teach simultaneously at several institutions, often rushing from town to town to do so. The scientific books produced during this period were preponderantly elementary textbooks, and the professional publications reflected the influx of newly trained scientists, often giving evidence of immaturity and provincialism. Scientific research was frequently also hampered by a lack of the simplest laboratory instruments. This resulted, in part, from official pressure on the academic laboratories and research institutes to prove their "usefulness," the demands being carried to the point of requiring them, at one time, not only to produce their own scientific instruments but to turn out machinery for industry. Little by little, these difficulties were overcome between 1920 and 1930. The quality of high-school education was restored, permitting the selection of well-qualified students for scientific study. The science faculties were augmented by an influx of competent graduates. Dependence upon imported scientific instruments was reduced by the creation of a domestic instrument-manufacturing industry, which was to develop so successfully that one of the things which strikes visiting observers today is the lavish equipment of leading Soviet laboratories.

But while the situation of Soviet science improved rapidly in respect of these material and human factors, another menace loomed in the growing pressure for ideological conformism. Suspicions had already been voiced in the early years of the regime regarding certain scientific theories that were viewed as heretical from the standpoint of dialectical materialism, but such attacks had been sporadic and ineffectual. Thus, the theory of relativity had come under fire, but physicists continued to teach and use it. The situation was less happy in biology because of its greater vulnerability to the inroads of extrascientific Marxist dogma. In fact, even before the revolution, certain trends had begun in Russian biology that helped prepare the way for its eventual emasculation under Stalin.

Among those originally responsible for injecting an ideological note into biology was Timiriazev, one of the most influential figures in modern Russian scientific development. An outstanding plant physiologist, Timiriazev was well ahead of his age in the application of physico-chemical principles to biology. But he was also an early devotee of Marxism and one of the few senior scientists who enthusiastically embraced the Bolshevik revolution. When he died soon afterward, he was enshrined as one of the revolution's patron saints, and even now Soviet scientists dare not admit that any of his theories was mistaken. For example, Timiriazev claimed that the absorption maximum in the spectrum of chlorophyll coincided with the energy maximum in the solar spectrum, and he held this alleged coincidence to be striking evidence of Darwinian adaptation of organisms to external conditions. Already in 1883 Engelmann had pointed out the error of Timiriazev's theory, but all Soviet treatises on the subject still hail it as a great "discovery" and proof of Timiriazev's genius.

In one region of biological theory, Soviet ideologists found themselves in agreement with the leading scientific school. Pavlov's theory of conditioned reflexes was considered to fit perfectly into the dialectical theory of the development of higher nervous activity in organisms in response to external challenges. The old master was therefore given generous support—and even the unique privilege of criticizing the official philosophy in his lectures.

In spite of the stagnation that affected many of the biological disciplines, the first fifteen years of the Communist regime were generally a period of rapid growth and expansion for Russian science. By 1930 scientists were well on their way to becoming a privileged class in Soviet society, in respect to not only economic status but also freedom to choose their field of work, to publish their research findings in Russian and foreign journals, and to maintain contacts with foreign scientists, including the privilege to travel and work abroad. All through the 1920's and early 1930's, Soviet scientists were permitted to spend months and even years working in the leading laboratories of Germany and England. Foreign scholars were invited to scientific conferences in the Soviet Union and lavishly entertained; some prominent ones went to the U.S.S.R. to work—for example, the American geneticists, H. J. Muller and R. Goldschmidt, who contributed a great deal to the development of a flourishing Russian school of genetics.

One of the Soviet scientists who worked abroad in this period was Peter Kapitsa, a promising young physicist. Taken to England by his professor to get over a crisis in his personal life, he was permitted to stay more than ten years at Cambridge, where he rapidly acquired an outstanding reputation in low-temperature research. Several times he made temporary trips back to the Soviet Union to attend scientific meetings; in 1935, during one of these visits, it was announced that he had decided to stay in Moscow. He was never again seen abroad, and the Soviet government later purchased his laboratory equipment in Cambridge so that he might continue his work in Moscow. Another outstanding young theoretical physicist of this period, George Gamov, managed to remain abroad permanently, first in Denmark and later in America, where he is now well known among scholars as an astrophysicist and to the public as a popular science writer.

Gamov's departure in 1933 was one of the signs that the honeymoon between Russian science and the Soviet regime was over. With the consolidation of Stalin's personal dictatorship, terror, which until then had been confined, more or less, to the political arena, invaded other areas as well. As in domestic and foreign policy, so in science and art, the personal whims, views, and phobias of the omnipotent dictator became decisive. In art, only things pleasurable to Stalin's eye or melodious to his ear were tolerated; in science, it was Stalin who decided whether Marr's theories of the origin of languages, or Mendel's laws of genetics, or Pauling's theories of resonance in chemistry were in agreement with dialectical materialism.

The best-known case history of Stalin's personal intervention in science was, of course, the destruction of genetics, with the role of executioner played by the self-made selectionist, Trofim Lysenko. The official motivation for Lysenko's campaign against scientific genetics was absurd. Briefly, genetics teaches that species can be changed only by selection of the few desirable mutations among many random ones; the number of mutations can be increased artificially, but it is (at least, as yet) impossible to direct them. Therefore, patient selection of progeny through many generations, ultimately producing genetically pure strains and crossing them to obtain the desirable combination of heritable characters, is needed to produce improved and hereditary stable breeds of plants and animals. These theories did not seem to fit in with the cardinal assumption of dialectical materialism, namely, that "being determines consciousness," that all phenomena in life can be explained purely in environmental terms. Nor did the principles of genetics fit in with the aims of a government which claimed to be able to change human nature by altering the economic and political structure of society. Finally, the theories of genetics seemed to imply that the improvement of agricultural plants and animals, so sorely needed by the Soviet state, could not be as rapid as its leaders desired. All of which made it easy for Lysenko to attack genetics as "fatalistic," contrary to the tenets of dialectical materialism and highly impractical to boot. According to Lysenko, heritable changes could be created to order, and on short notice, by the changed feeding of plants or animals, by grafts, by selection of individual seeds, or by seed treatment prior to germination. Genetics scoffed at his claims—so genetics had to be suppressed.

Lysenko's campaign reached its climax in August, 1948, during a "scholarly" meeting of Soviet biologists, at which he dramatically announced that his theories had been "examined and approved" by the Central Committee of the Communist party—in other words, by Stalin himself. An orgy of persecution against "academic" geneticists was soon unleashed. Institutes of genetics were closed; textbooks were destroyed; professors disappeared from their laboratories. Lysenko was made president of the Lenin Agricultural Academy, founded by Lenin on the insistence of the leading Russian geneticist, N. I. Vavilov. (Vavilov himself had been attacked by Lysenko as early as 1940. Before the beginning of the German-Soviet war, he was accused of planning to flee abroad and was deported to Magadan in northeast Siberia, where he died in January, 1943.) Within a short time, the young but flourishing school of Russian genetics was totally destroyed.

At the present time, several years after Stalin's death, the status of genetics in the U.S.S.R. is still obscure. Lysenko has been demoted from the presidency of the Agricultural Academy, and his theories have been subjected, now and again, to critical comment by bona fide scientists. Yet, Khrushchev has on several occasions gone out of his way to praise Lysenko's abilities, and as recently as December 8, 1957, *Izvestia* carried Lysenko's article entitled "Theoretical Successes of Agronomic Biology," which boldly repeated all his former claims and prevarications. Whatever Lysenko's personal position at the present time, the fact remains that the Russian geneticists who survived the reign of terror have given scant indication of having resumed their work.

The saddest aspect of the persecution of genetics was that distinguished scientists were used as tools of Stalin's purge. The president of the Soviet Academy of Sciences at that time was S. I. Vavilov, younger brother of the geneticist and himself an outstanding physicist, with considerable achievements in physical and physiological optics. A liberal European in his philosophy, an admirer of his older brother, he had to sign orders for the destruction of his brother's life work and the purge of his pupils. He had to praise publicly the quackery of Lysenko and the scientific genius of Lysenko's patron in the Kremlin. Western scientists are apt to ask about Vavilov and others like him: If they could not rebel, could they not at least resign? Was it only fear of the dictator that caused them to deny their principles and to forget their personal attachments? Or were the glitter and economic rewards of the presidency of the Soviet Academy sufficient to buy a man's conscience? Or did the prospect of leadership in the broad expansion of scientific work for the benefit of one's country lead to a rationalization which made capitulation in any single area appear justified? Was solace provided by the hope that, by remaining in an exposed position and offering oneself as willing executioner of the dictator's orders, one could protect other victims and thus soften the blows? Would it not have meant opening the door to greater evil to withdraw and let the presidency of the Academy be taken over by an adventurer like Lysenko—perhaps, Lysenko himself? It is difficult for those who did not experience the debilitating pressure and the moral dilemmas of life under terror to comprehend the rationalizations (and self-deceptions) which permit men to carry on under these conditions. We cannot but wonder how intellectual and scientific life could continue at all and bear fruit in daily expectation of arrest, deportation, and death, in constant dissimulation of one's true feelings and convictions.

The historic fact is that Soviet science *did* function and bear fruit even during the darkest days of Stalin's terror. Of course, genetics was obliterated. The blight soon spread to other branches of biology and, for a while, invaded some sectors of physics and chemistry. From his vantage point as president of the Agricultural Academy, member of the Academy of Sciences, director of a great research institute, boss of all agricultural research stations and protégé of the almighty dictator, Lysenko threatened to become the master of biology (if not of all science) in the U.S.S.R. Distinguished physiologists and embryologists—Orbeli, Shmalgauzen, and others—were censured and made to recant in public. Strict "Pavlovism" was proclaimed obligatory in the physiology of the nervous system and in psychology. The quackery of Madame Lepeshinskaya, who claimed to have proved the existence of extracellular "living matter," was rewarded with a Stalin prize.

In physics, the most persistent attack was directed against the statistical interpretation of quantum mechanics—the indeterminacy principle of Heisenberg, the complementarity concept of Bohr. The tendency of many modern theoretical physicists to discard the concept of "reality" of physical phenomena and replace it by a system of mathematical relations between "observables" was a challenge to dialectic materialism, which is based on a belief in the inherent capacity of the human mind to perceive the realities of the physical world (and on the narrow nineteenth-century postulate that these realities must consist entirely of different "modes of motion" of matter). Many physicists in the West, including Einstein, De Broglie, and Schroedinger, refused to accept statistical "laws," predicting, within definite limits, the results of practically possible experiments, as an adequate substitute for causal laws connecting real phenomena of nature, on an atomic scale as well as an everyday scale. No wonder that the search for a reinterpretation of quantum mechanics on a strictly causal basis was, and still is, officially encouraged in the Soviet Union.

However, no sustained effort was made to destroy those sharing the prevailing Western ideas in quantum mechanics as thoroughly as the geneticists were destroyed in 1948. Rejection of the "idealistic Copenhagen school of theoretical physics" (Bohr, Heisenberg, Born, and the majority of Western theoretical physicists) became obligatory and may still be so. Some venerable Soviet physicists, such as Joffe and Frenkel, were taken to task for being lukewarm in their conversion to dialectic materialism, applying it only in the introductory chapters of their books in the hope that censors would not read farther. They had to publish recantations and promise better behavior in the future, but the threat to them, though ominous for a while, has subsided in recent years.

The purges, however, did not entirely spare Russian physicists. German-Jewish physicists, who went to Russia after Hitler's ascension to power, were summarily expelled after the Soviet-Nazi rapprochement in 1939; some of them (like Houtermans) were arrested and kept imprisoned for many years. The most brilliant theoretical physicist of the new Soviet generation of scientists, Landau, also went to prison. Khrushchev himself has reportedly stated that the dean of Soviet experimental physicists, Joffe, was on the list of people to be shot in one of the purges but that his name was crossed out by Stalin because "he may still prove useful to us." According to apparently reliable reports, Kapitsa was punished—allegedly for refusing to devote his Institute to work on the atomic bomb—by spending several years under house arrest in the country.

One reason why physics and chemistry have, on the whole, been treated much more gently than biology may have been the vital interest of the Soviet leaders, including Stalin himself, in the contribution these disciplines could make to the military and industrial might of the Soviet Union. They were the hens expected to lay golden eggs and therefore not to be killed. And useful they did prove. Spurred by material encouragement, by pressure from the political leadership, and by patriotic spirit (which has always been a strong impetus to Russian science, especially during the German invasion of 1942–45), Soviet physicists rapidly mastered the new techniques of nuclear physics. They produced the first Soviet atom bomb in 1949, only four years after Hiroshima. They arrived, almost simultaneously with American scientists, at a practical solution of the problem of thermonuclear explosions. They now hold the lead in the construction of great particle accelerators. Finally, continuing the tradition of Russian pioneering in aerodynamics and ballistics, they stayed well ahead of American development in rocketry and produced, in 1957, the first man-made satellites and the first intercontinental missiles. With this last success, Soviet science has emerged as a great international force, its prestige enormously enhanced. Speculations are rife as to what influence this new force will exert on Soviet society and on the future world policy of the Soviet Union.

The idea that the Soviet Union may be a hell politically but a paradise for scientists (at least in some areas) is difficult to accept.

Some scientists in the West took this view in the 1920's, but as more and more sciences and scientists came under Moscow's fire, as Lysenko trampled over genetics, and as quantum mechanics was bombarded by the heavy guns of dialectical materialism, even these Western observers began to think that Soviet science faced a hopeless future, that ideological perversion would eventually cripple it, and that freedom of science can never be secure in a country which has no respect for freedom of thought in general. The material and societal inducements offered by the totalitarian dictatorship of the East appeared to be inadequate substitutes for independence of thought and personal freedom offered by the libertarian societies of the West, even if the latter were much less actively interested in science and much stingier in its support.

However, the discounting of Soviet science in the 1940's was as exaggerated as had been its exaltation in the 1920's. A climate of spiritual freedom is certainly important for the development of science, but so are adequate education and proper support. Although the Soviet Union was deficient in the first, it provided plenty of the other two. Discussing the state of Russian science in 1952, during the period of Lysenko's ascendancy, this author wrote:

> One is left with the impression of a flourishing community stricken by an epidemic. Here and there, houses are closed and quarantined. Nobody knows for certain how to protect himself from the scourge. A certain ritual of pacifying the angry gods is followed. It includes praising their wisdom at the beginning and end of each printed or spoken communication and loudly denouncing all those stricken by the plague as sinners who have brought it upon themselves. . . . However, as men do in real epidemics—as they did during the war under the bombs—those who are still hale go to work as usual.

And in conclusion:

> It is wrong to think of contemporary Soviet science as being largely paralyzed by the ideological dictatorship of ignorant politicians. Some branches may be dead or stunted, but it is still a vigorous, growing tree. Limbs that have been cut off sprout new shoots; those prevented from growing in their natural direction, grow around the obstacle.[1]

There is not much to change in this appraisal now, except that the situation has improved considerably since Stalin's death. The areas of ideological oppression have narrowed, and its intensity decreased. Not that Soviet science is now quite free; far from it. Occasional pressure is still exerted to restrict activity to immediately practical ends; no group of scientists is quite immune from censure

[1] *Bulletin of the Atomic Scientists,* August, 1952.

for doing things which are deemed of no use to the state. Lysenko continues in the picture as research director and editor of "agri-biological" journals. When Nikita Khrushchev recently proclaimed that the Russian dairy industry would overcome the American lead within a few years, his optimism was based on Lysenko's promise that Russian cows can be induced to double their production of milk—one should add, by means as "scientific" as those which he once promised would transform agricultural plants. Physicists using the principles of relativity, uncertainty, or complementarity are still exposed to ideological heckling. Obeisance to dialectic materialism as the mainspring of all true science remains obligatory. But science, which has weathered the days of Zhdanov and Stalin, will not be stunted by today's much milder forms of control.

A point to be underscored is that science in Russia, even under worst conditions, has been freer than all other areas of intellectual endeavor. Compared to historians and sociologists, economists and writers, who have lived in constant jeopardy, trying to guess what tomorow's party line would be, recanting today what they said yesterday, scientists have been much more their own masters. This has made scientific careers particularly attractive to those among the younger generation who most value their independence. Thus it is erroneous either to compare or to contrast the pressures and restrictions which have impaired the attractiveness of science in the West with the much greater pressures and tighter restrictions in the Soviet system. In America, even a little pressure can cause the most valuable young intellects to seek expression in other fields. In the Soviet Union, much heavier restrictions have not prevented science from seeming a haven of intellectual freedom.

It is difficult to decide whether general restrictions on independent thinking and the training of Soviet youth for conformity discourage to any extent truly creative scientific thought (as contrasted with good scientific workmanship). Scientists with revolutionary new ideas are rare under all conditions; it is difficult to apply statistics to such essentially unique phenomena. Does systematic indoctrination in philosophic dogmatism and political docility—the basic feature of instruction in all Soviet schools and colleges—make the appearance of a Darwin, an Einstein, or a Lobachevsky less likely? Despite the vast quantitative growth and steadily improving average level of scientific achievement in the Soviet Union, so far, at least, no great new ideas have come from there. Since 1904, when Pavlov won a Nobel award, this prize has gone only once to a Russian scientist: the chemistry prize awarded to N. N. Semenov for his work on chemical chain reactions—a highly creditable but not revolutionary achievement. In the same period, over twenty Nobel prizes in science have gone to scholars in the United States. The majority of them were for achievements similar to that of Semenov, but in a few cases revolutionary new concepts were involved. An outstanding example was last year's award of the physics prize to two young Chinese-American physicists, Yang and Lee, for the overthrow of the parity principle.

Is a similar revolutionary idea less likely to occur to a Soviet scientist because he has been discouraged early in his career from questioning the prevailing concepts? Or will pressure for conformity merely provoke the keenest minds to doubt? Some of the greatest scientific geniuses—Copernicus, Galileo, Kepler—conceived their heterodox ideas despite prevailing pressures to conform. Some competent Soviet scientists seem to have accepted the tenets of dialectic materialism as a matter of conviction and not merely through compunction or convenience; yet it seems a fair guess that there are plenty of Russian scientists—old and young ones alike—who see in the official philosophy only cant, to be used occasionally for self-protection.

Whether or not the general discouragement of free inquiry in Soviet academic and public life also inhibits the inquiring mind in science, there is no doubt that other forces at work in the Soviet Union are powerfully assisting scientific progress. In the first place, Soviet secondary education introduces all youth to the ideas and techniques of mathematics and science. The curriculum itself—except for the emphasis on ideological studies and the disappearance of classical languages—is not very different from that offered by high schools in prerevolutionary Russia (or, for that matter, in many other European countries). What is, of course, radically different from prerevolutionary days is the enormous spread of secondary education and the opportunity given to gifted children to qualify for further training in scientific careers.

Two other factors favoring the development of science are more uniquely typical of Soviet society. The first is the prestige with which science is endowed, which finds expression both in the financial rewards of a successful scientific career and in the high standing of scientists in public opinion. The second is the lavish support the government gives to institutions of higher learning and to research institutes. The facts relevant to these considerations have been widely discussed since the launching of the first satellite. They include the larger numbers of scientific and engineering graduates in the U.S.S.R. compared with the United States; the relatively high salaries and other benefits in science; the conviction of Soviet scientists that "money is there for the asking"—so different from the often precarious financial status of the most important research projects in the United States.

There is another factor which deserves more attention than it has received from the West: that is the Soviet glorification of science as a way to achieve happiness and prosperity for the peoples of the U.S.S.R. and the whole world. This faith in science as a creative force in society is, perhaps, even more important in attracting the best minds to science than the expectation of financial rewards and social recognition. Soviet youth grows up in a world of poverty and privation. It is not hard to instil in them the desire to transform society, through science, into something much happier and more prosperous. Small wonder, then, that in publicizing the aims of scientific development, the regime puts less emphasis on the enhancement of Soviet military might than on more attractive and appealing aspects, including exploration of space and interplanetary travel, which are calculated to arouse the enthusiasm of youth.

While world attention remains glued on Soviet successes in producing thermonuclear weapons, ballistic missiles, and space satellites, there is a tendency to forget that this is only one aspect of the broad development of Soviet science. The regime aims at making the U.S.S.R., through science and technology, not only the most powerful but also the most productive country in the world and to induce other nations to accept its technical and scientific help and leadership. It is to this wider aim that the enormous organization of scientific research in Russia is directed.

The direction of this effort is centered in the Soviet Academy of Sciences. In contrast to similar institutions in the United States and Western Europe, the Soviet Academy is not merely the leading learned society in the country; it is also the operating agency for much of the research done in the U.S.S.R. According to a recent study by A. Vucinich, of Stanford University, the Academy now has one hundred eighteen institutes in the various natural sciences. Besides sixteen local branches of the central academy, there are twelve affiliated academies in the constituent republics of the U.S.S.R.

In viewing the work of this vast institute, Western scholars are inclined to note primarily its subordination to party control and to the political aims of the regime. If they are sociologists (like Professor Vucinich), they are impressed by the low scholarly level of the historical, sociological, and economic sections of the Academy and its subservience in all these fields to the current party line. Attention is called to the hollowness of the autonomy of the Academy, the powerlessness of its supposedly supreme General Assembly, and the influence of its party-dominated Presidium and permanent secretariat.

All this is undoubtedly true. However, in most of its important *scientific* functions, the Academy is a highly competent and largely independent body. It is composed of a fair selection of the best minds in Soviet science, including a considerable number of older men with prerevolutionary education. Those who owe their membership to considerations other than scientific prominence (such as Lysenko) are small in number. Although Nesmeyanov, the present head of the Academy Presidium, and Topchiev, the chief scientific secretary, are both Communist party members, they are also prominent scientists, not party hacks or bureaucrats.

The Academy has been described as the general staff of a scientific army, throwing its columns to this or that decisive segment of the scientific "front" in response to the directives of the Communist leadership; in fact, this is the picture that the Presidium itself likes to paint in official reports. Science is visualized as a centrally planned activity; its most spectacular achievements are interpreted as results of planned concentration on a few "targets." This is only partly true. Efforts in applied science can be planned; if considered of national importance, they can be speeded up by a crash program. The United States' "Manhattan Project," which produced the atom bomb in 1945, is an example. It is quite likely that the Soviet space satellites were the result of a similar crash program.

The advance of fundamental science is much more difficult to plan. It has been reported that when the requirement of planning was first imposed on the research institutes the directors presented, as the "plan of research for the next year," the actual results of the previous year's research, in order to assure 100 per cent fulfilment. As in other fields, probably a good deal of this type of deceptive manipulation of the plan goes on, but this is not to say that central planning is totally ineffectual. To cite an example, in 1946 a conference was called by the Academy to review the state of photosynthesis research; it was found to be backward, and a decision was taken to promote it. By 1957, a second conference on photosynthesis saw the presentation of about one hundred fifty research papers as against ten at the earlier gathering. In a similar way, research in electronic computers and automation was spurred by resolutions of the Academy.

However, pressures of this type are much more likely to spur progress in fields where Soviet science is recognized as backward compared with the West than in forays into new, unexplored regions. They can bring up the rear but not lead the van. Resolutions proclaiming the necessity of achieving breakthroughs into unexplored areas of physics or biology are bound to remain mere words until a creative scientific mind shows the way, usually in an unexpected direction.

All attempts at central planning notwithstanding, the decisive determinant of new achievements in science remains the personal interest and dedication of individual scientists; and this cannot be planned. The recent development of areas of science which require expensive instrumentation and the co-operation of large groups of researchers has not invalidated the fact that science is what individual scientists make it. Particle accelerators were developed in America, not because a central agency had planned it, but because of Ernest Lawrence; Soviet science owes its success in the same area largely to Wechsler. In short, the development of fundamental science in Soviet Russia proceeds in about the same way as in the West—through the interplay of individual talents.

As for the alleged "unevenness" of scientific progress in the Soviet Union, attributed to planned concentration on a few sectors, this again, if true at all, affects *applied* science only. If one takes wide classifications (such as "nuclear physics," "molecular physics," "atomic physics," "organic chemistry," "enzymology," and so on) and counts the numbers of papers published in these classes in the U.S.S.R. and in the United States, one finds a more or less constant ratio, indicating that Soviet science—in-so-far as broadness of coverage is concerned—has generally caught up with Western science. One does find wide differences in special areas, but differences are easy to find in the scientific record of every country. What is significant is that this kind of statistics, extended over several recent years, reveals a continuous increase in the relative contribution of Soviet research to most if not all areas of science—an increase that will probably become more and more conspicuous in the years to come.

However, scientific leadership does not consist in numbers—be it numbers of scientists or of scientific papers; it is based, above all, on great individual achievements (Holland and Denmark, for example, have played a role in science out of all proportion to the numbers not only of their populations but also of their scientists and scientific publications). Whether such achievements will come in increasing numbers from the U.S.S.R. because of the growth of its total scientific manpower and resources, or whether they will remain disproportionately small because of the general prevalence of conformism and the discouragement of unorthodox thinking, only time can tell.

Stop before Turning

Looking through the window of our conference room and seeing skiers dashing down the slope, gathering speed, it occurs to me that the nations of the world are rushing down in a similar way, toward an unknown abyss. Their first problem is how to bring this perilous dash to a stop so that they can start climbing toward the summit of stable peace. Since history provides no chair lift or T-bar, this ascent is bound to be slow and painful. This paper deals with the first step of this program.

Dr. Jerome B. Wiesner postulated in his paper that in order to make possible the first small steps toward secure peace a long-term program should be outlined, at least roughly, at the very beginning. He argued that, in this way, reasonable hope for subsequent, more substantial agreements will make more attractive the earlier and, on the face of it, not very engaging ones. I agree with him.

Dr. C. H. Waddington suggested—and I agree with him, too—that the most promising openings are likely to be found not in the field of disarmament (or other agreements on mutual restraint) but in positive co-operation in scientific, technological, or economic areas. Mutual trust is what we need (as Dr. A. V. Topchiev has repeatedly pointed out), and it is best achieved by co-operation in undertakings broadening the area of common interest. In this way, the issues which divide the world and cause mutual apprehension among nations could be deprived of the dominant role they now play in the minds of the peoples of the world.

I believe that Dr. Morton Grodzins was also right in reminding us that disarmament is more likely to follow than to precede political settlements because the existence of festering political sores—areas where conflict is ever ready to break into the open—will not permit the growth of the mutual trust needed for properly safeguarded disarmament and the establishment of a world security system, as advocated by Mr. Richard S. Leghorn.

Lastly, I believe that a very real factor in the distrust between West and East lies in two different (although not necessarily contradictory) faiths for which each side claims universal validity. Assertions by each side that it has no intention of spreading its respec-

Paper presented at the second COSWA-Pugwash Conference, at Lac Beaufort, Quebec, April, 1958, and later printed in the *Bulletin of the Atomic Scientists*, September, 1958.

tive creed by force are not enough in the face of their belief in the ultimate and complete triumph of their cause, as epitomized by Mr. Khrushchev's terse phrase, "We will bury you." The imposition of the Communist system on several East European countries by the pressure of military occupation cannot be forgotten.

It follows from all this that a long-term program of approach to stable peace, such as Dr. Wiesner proposed, cannot consist merely of successive disarmament steps, each made possible by the increase in trust following the implementation of the preceding step, while political agreements are postponed, and ideological antagonism is left to take care of itself. Rather, simultaneous advance must be planned on a wide front, including military, political, economic, and ideological sectors. Truly effective disarmament steps may have to be relegated to relatively late stages of the program, since these steps—as contrasted with those merely slowing down the arms race, such as a bomb test moratorium—require verification mechanisms which would be difficult to implement until the political climate of the world is radically changed. If the necessity for a broad simultaneous advance is accepted, the first step in each field will have to be modest. It must aim, in my opinion, at the *stabilization of the existing situation.*

POLITICAL STABILIZATION

Stabilization of the present political map of the world will imply many painful decisions, particularly for the West. Because of the wide advance of Soviet power after the war, in Europe and Asia, the West now bases many hopes on at least partial reversal of this advance and therefore postpones the full legal or political acknowledgment of the changes. The Soviet Union, on the other hand, can be satisfied with the present distribution of political power in Europe and should be willing to pay for its stabilization with the renunciation of further advances in areas such as the Middle and the Far East, where such advances seem possible.

It would be difficult to reach specific agreements on many separate problems of political stabilization (as suggested, *e.g.*, by Mr. Kennan), except perhaps after very prolonged and uncertain negotiations, similar to those over the Austrian peace treaty and the Korean armistice. It may be equally, if not more, difficult to balance a disadvantageous settlement of one political controversy against the advantageous settlement of another one to obtain a mutually acceptable "package." Perhaps it would be easier, faster, and more effective to try to agree first on the general principle of recognizing all *de facto* political divisions existing in the world. A common declaration of the major powers' intention of accepting this principle would in itself be a very effective means of allaying the present uneasy state of mind.

This decision will be painful because it will mean writing off, if not permanently, at least for an indefinite length of time, many cherished—and among them, some legitimate—hopes kept alive by the American policy of non-recognition of the postwar changes in Eastern Europe. That we in fact do not intend to assist actively in the restoration of the freedom of the nations of Eastern Europe—as the "rolling back" promises of the 1952 presidential campaign seemed to imply—was demonstrated by Western inaction at the time of the Hungarian uprising. The hope that "next time" the West will intervene more effectively is unfounded and dangerous (like the hope widely held in America after the Korean War that "next time we are not going to fight with one hand tied behind our back").

The recognition of the *status quo* in Europe will have to embrace first of all the present German-Polish border along the Neisse and Oder rivers. This will remove one of the strongest sources of apprehension in Poland and other Eastern European countries and lessen their dependence on the Soviet Union for their existence in the established geographical framework. The hope for these countries to gain more internal freedom and to resume friendly relations with the countries of the West now rests in the gradual relaxation of the rigidness of the Soviet policy. This relaxation is likely to be favored by the removal of the apprehension that the West will support a renewed German expansion toward the east.

American recognition of the *status quo* in eastern Europe is likely to produce a feeling of letdown among our east European friends. But do we really serve their cause by sanctimonious non-recognition of the *de facto* state of affairs in their countries and by appeals for resistance which we are unwilling to back up in the crucial moment?

The most difficult aspect of the stabilization problem in Europe is clearly Germany. If it were possible to avoid "freezing" its division into two parts by a prompt agreement on neutralization and reunification, this solution should be attempted. Whatever procedure is used for reunification, it now seems likely to end in the disappearance of the Eastern regime rather than in the spread of communism to Western Germany because of the greater size and power of the latter and the greater support of the Western regime by its population. Therefore, there seems to be no reason for the West to oppose reunification by *any* peaceful method, including negotiations between the two regimes leading to a confederation, as suggested by the Soviet Union. Unfortuantely, it seems unlikely that unification of a neutralized Germany can be brought about rapidly enough to take precedence over the stabilization of the European political map as a whole. If this is true, stabilization should be attempted without further delay.

No judgment on the historical or ethnical justice of the present political division of central and eastern Europe is implied in these decisions, but only sober recognition that from now on any attempt to change this division by force is too dangerous for world peace to be contemplated with equanimity—let alone encouraged, as it is by our present policy of indefinite non-recognition. The argument that our disapproval of the *status quo* (or formal lack of political recognition, as in the case of the Baltic republics) undermines the hold of Communist governments on these countries and sustains our moral authority in the world has repeatedly failed to prove its validity. Non-recognition of the Communist government of China is the most flagrant example of the utter ineffectiveness of attempts to sustain a "moral" position in world affairs without being prepared to act upon it in any practical political or military way.

The importance traditionally attached to recognition in American foreign policy will make it difficult for the American government to extricate itself from its present untenable position without creating the impression of endorsing the foreign rule imposed on previously independent countries and betraying the traditional American role as supporter of small nations in their fight for national independence and political freedom. However, so many holes have already been punched in this mantle that the king has become practically naked. Acknowledging that the present political map of the world cannot be safely changed by force or external pressure does not mean betraying the faith, shared by most democratic societies, that sooner or later political freedom and the rights of the individual will triumph everywhere in the world—any more than a similar renunciation of active support of Communist activities abroad by Soviet Russia would mean betraying their faith that one day the Communist economic system will be accepted by all mankind. What is important is that the recognition of the existing political boundaries should be given by both sides, not grudgingly, as in a military truce, but wholeheartedly, in recognition of compelling necessity to establish a base line for progress toward a more equitable and stable peace.

The same stabilization principle will have to be applied also to the divided countries in the East, Korea and Vietnam. Their unification prior to the freezing of the *status quo* would be even more difficult to achieve than that of Germany because in these countries both regimes appear to have intense native support, as shown in their civil wars. Stabilization will have to mean also renunciation of the conquest of Taiwan, and liquidation of the pretense that the Taiwan regime should be considered—and act—as the central government of all China. That mainland China will have to be "recognized" and admitted to the United Nations hardly needs saying.

In the Middle East, and perhaps in some other parts of the world as well, stabilization of the *status quo* may have to be supervised by international surveillance, like the United Nations patrol on the Israeli-Egyptian frontier after the Sinai war. As part of the stabilization agreement, the Soviet Union will have to recognize the *de facto* borders of Israel, thus sacrificing one of its present claims to Arab sympathy. As in Europe, this frontier will have to be stabilized irrespective of the historical or ethnical justice of the rival claims, simply because it exists and cannot be moved without violence.

Many formerly colonial areas are in turmoil and changes in them are inevitable. A temptation exists for major nations to support different factions or even to foment their struggles. If this support were withdrawn and internal forces left to assert themselves, internal revolutions, as well as fusions and fissions on the political map of Asia or Africa, could be contemplated with relative equanimity by all major nations. A useful step in the presently embattled area might be an effective embargo on arms delivery, as suggested by the Soviet Union. As in other regions, a more significant forward step could be active co-operation rather than mere mutual restraint. If the major powers could undertake a common effort for the economic and technological rehabilitation of the Middle East, their interests in this area would be, at least in part, merged. Mr. Adlai Stevenson suggested in one of his speeches that since neither the United States nor the U.S.S.R. was willing (or able) to build the Aswan dam by themselves, perhaps they could build it together. A more general proposal for such a common effort was made by Italy.

Military Stabilization

The arms race itself offers a typical example of possible stabilization through cessation of nuclear arms tests. The significance of this step in itself is relatively small, and the high hopes which a large part of world public opinion has set on it seem to me pathetically exaggerated. To believe that a test ban will open the way for truly significant progress in the elimination of atomic weapons is totally unfounded optimism. However, the test ban can acquire significance if, instead of being considered the beginning of a linear progression of disarmament steps, it is looked upon as one of the measures implementing a broad stabilization of the world *status quo*. The cessation of tests, although it will not freeze the arms race as a whole, will significantly slow down its pace. It will freeze the state of one particularly alarming military art and make it more difficult for new nations to obtain atomic weapons. (For the latter purpose, the stabilization agreement will also have to include provisions for not supplying nuclear weapons, materials, or weapons know-how to nations not now possessing them.)

The reason that an agreed cessation of tests is feasible is that the kind of monitoring and inspection needed to control compliance with this agreement does not exceed what is possible, even in the present climate of international distrust. An important factor is that the risk of a single, or a few, evasions of the ban could not have consequences decisive enough to make the risk of illegal tests worth taking; by the same token, other parties to the agreement need not feel too greatly endangered by the possibility of a few evasions. In other words, a certain degree of uncertainty of detection can be accepted in this case as a calculated risk.

There may be other components of the arms race amenable to stabilization at the present level without requiring more radical forms of international control than the test ban. Perhaps a committee of experts in the United Nations could be formed to seek out such areas and propose methods of stabilization applicable to them. Particularly important—if technically possible—could be the freezing at its present level of the development of long-range military rockets, perhaps by the establishment of a joint international space flight agency. The reasons for the rejection by the Soviet Union a few years ago of a similar UN plan for atomic energy do not apply now. One reason was the possible economic importance of the development of atomic energy. This was quoted by the Soviet Union to explain its reluctance to place this development in the hands of an international agency. The other, unadmitted reason was the inferiority of the Soviet Union in atomic technology at the time when the plan was proposed and the apprehension that this inferiority would be perpetuated by its acceptance. At the present time, the freezing of military rocket development and the transfer of nonmilitary space flight development to an international agency could perpetuate, if anything, an advantage of the Soviet Union; and the economic importance of space travel does not appear real enough for its internationalization to arouse any apprehension.

MITIGATION OF IDEOLOGICAL CONFLICT

Even if the fears engendered by the atomic arms race and the present feeling of political instability in many parts of the world could be removed, by slowing down or stopping the race and stabilizing the political map, a deeper source of distrust would remain. This is the already mentioned dedication of the two great camps to two ideals which, although not obviously contradictory, claim such complete allegiance as to appear mutually exclusive. One is the Communist economic system, which its adherents believe to be a superior form of society that all mankind is bound to choose sooner or later. The other is political freedom, which its adherents believe to be the highest development in the political history of mankind, which all nations will adopt sooner or later. However strongly the adherents of the two creeds may proclaim their intention not to use force in assisting in the "inevitable" spread of their creed, the inherent dynamism of both ideas will make the growth of mutual trust difficult. A gradual subsiding of the intensity of the younger (and therefore more violent) creed may reduce this danger in time, as aggressive proselytizing for new religions has been gradually reduced in violence in the past. However, those who already discount the role of Communist ideology in inspiring the political action of the Soviet Union, as Bertrand Russell does, seem to me to run far ahead of reality.

The agreements on stabilization of the political state of the world should therefore be supplemented by provisions for keeping in check the dangers arising from the ideological conflict. This is a difficult problem, and only partial solutions of it are possible. Freedom of thought and freedom of expression are among the most cherished rights of free men. They cannot be restrained by an agreement entered into by any democratic government, however irritating this may be for other nations. All that the American government could undertake would be to show an example to its people by refraining from unbridled attacks on the institutions, beliefs, or personalities in the other camp and from exhortation to resistance or revolt. It could impose some discipline on the public utterances of its high officials (as long as they are in office), although in America, as contrasted to some other free countries, even the highest government officials claim for themselves full freedom of public comment.

The situation is quite different in the Communist camp, where government exercises effective control over all public statements and can turn off or on the expressions of public approval or indignation like water in the bathroom tap. On the other hand, opposition to the capitalist economic system is a much more fundamental pillar of communism than the opposition to communism is in the American view of the world. To stop attacking capitalism is more than one can ask of the Communist leadership or its official press. All we could hope is that these attacks could be made less virulent and more objective. After all, the Soviet theory of inevitable historical conflict between the obsolete capitalist and progressive Communist systems by its very determinism seems to make attacks on the motives and intentions of the representatives of the obsolete order unnecessary.

Responsibilities of Scientists in the Atomic Age

Everybody has a responsibility to the society of which he is a part and, through this society, to mankind. In addition to the common responsibility of all citizens—such as, to obey laws or to pay taxes—many individuals have additional responsibilities arising from their belonging to special groups that are endowed with special capacities, possess special knowledge, or enjoy special power. The doctor, the teacher, the minister, the policeman, the soldier, have such special responsibilities, often covering not only their work or service but their behavior in general. In our time, when science has become an important force affecting both the life of the individual and the fate of society, scientists have acquired a peculiar responsibility originating from their special knowledge and the power associated with it. What is it?

More clearly than anybody else, scientists see the senselessness and the tragedy of the present situation of mankind—the *reductio ad absurdum* by modern technology of the historical tradition of humanity divided into warring factions which threaten each other with armed might. Despite this knowledge, scientists remain the weaponeers of hostile nations. They are caught in a vicious circle. If scientists refuse to provide their country with all the weapons science can invent, they may be responsible for putting their nation at the mercy of another, hostile one. This was the motivation which caused the greatest physicists of our time to urge upon the American government the development of the atom bomb at the beginning of the Second World War. Without the fear that Germany might be the first to threaten the world with an atom bomb, scientists in England and America would not have mustered up sufficient enthusiasm for the job and the feeling of urgency which made the atom bomb a reality before the end of the war. After the war, it was apprehension—and as it later transpired, a justified apprehension—that the Soviet Union might be close to the production of a thermonuclear bomb that caused a number of American scientists to press for the development of the American H-bomb.

Paper presented at the third COSWA-Pugwash Conference, in Kitzbühel, Austria, in August, 1959, and later printed in the *Bulletin of the Atomic Scientists,* January, 1959.

Some people (scientists as well as non-scientists) see the root of the tragedy in wrong personal decisions of scientists. Max Born has lamented that his former pupils and co-workers—Heisenberg, Oppenheimer, Fermi, Teller, among them—have not learned, as he himself did, the wisdom of not lending their genius to the evil purposes of weaponeering. Robert Jungk, in his book *Brighter Than a Thousand Suns,* presented the whole history of the atom- and hydrogen-bomb development as one of the failure of scientists to make the correct moral decision, which was to refuse to supply their governments with the terrible new weapons. These critics do not admit that a conflict exists between two ethical imperatives with which a scientist is confronted in our divided world—the conflict between the voice of scientific conscience, which counsels him against putting his knowledge and skill in the service of destruction, and the counsel of loyalty to his state and his society, which tells him that he is not entitled to decide on his own whether this country and this society should be left to face an enemy with inferior weapons (which is likely to mean, in effect, with practically no weapons at all except moral strength and consciousness of rectitude). Some—a relatively small number—of scientists escape the conflict by becoming "conscientious objectors." This is the credo of the Society for Social Responsibility of Scientists. Among its members are some prominent scientists—Max Born, Kathleen Lonsdale. We have been told—even if we do not know it for certain—that, in the Soviet Union, Peter Kapitsa has refused to put his capacities in the service of atom-bomb development. It is, however, unlikely that, in any one country, a majority of scientists will choose this path and thus effectively impose unilateral disarmament on their own nation.

Dr. Jungk suggests that it may have been only lack of communication that prevented an agreement between scientists on both sides of the Second World War not to make an atom bomb, which would have enabled both sides to escape the dilemma of unilateral disarmament. According to his story, leading German nuclear physicists consciously and consistently evaded working in this direction and misled the German government as to the possibility of the development of the bomb. If only these German scientists had made a determined effort to let their American and English colleagues know of their decision! Then, Dr. Jungk suggests, Western scientists could have followed the example of their German colleagues, and the American A-bomb project would have bogged down, as the German one did.

I believe that the conspiracy of German scientists not to give the atomic bomb to Hitler, as described by Jungk, is a *post factum* rationalization of a vague uneasiness which had caused German scientists to drag their feet rather than put into the atomic bomb work the same enthusiasm and urgency that animated their British and American colleagues. The German scientists, while not outright defeatists, had not the same fear of Hitler's defeat as the Western scientists had of a German victory. They grasped at evidence that the bomb could not be made in time for use in the war; and their ascientific, if not antiscientific, officialdom did little to urge them on.

I believe that the basic cause of the predicament, into which the discovery of nuclear energy has brought the world, lies not in the inadequate ethical standards of scientists or in the difficulties of communication between them. Rather, it lies in the low ethical standards of national governments and the difficulties of communication between them; and this, in turn, is the consequence of the stubborn survival of an obsolete organization of mankind, its division into separate sectors which require—and receive—full and exclusive loyalty from their members. Within such a world system, only a few individuals are likely to claim and assert the right to hold to their self-set standards of moral behavior, when this behavior—if emulated by a large part of their colleagues—could inflict fatal damage on the society to which they belong and to which they owe their freedom and their living. Scientists as a profession are not likely to assume this attitude, any more than a whole age-class of American draftees is likely to turn one day into conscientious objectors. Scientists may be intellectually more advanced than the average of the population; but this intellectual level has nothing to do with attitudes in the face of this dilemma. While scientists see more clearly than others the terrible consequences of the use of the weapons they are developing, they see with equal clarity the possible consequences of their nation's being left at the mercy of an enemy equipped with them; they are less likely to cherish the illusion that "old-fashioned" weapons can provide adequate defense against modern technological armaments or that these armaments may not be brought to bear on the decision in a major future war.

The conflict between loyalty to his scientific ethics and loyalty to his nation weighs as heavily on the conscience of an atomic physicist as it does on that of a farmer, worker, or lawyer called into military service. It is not because of insufficient moral fiber, or indifference to human suffering, or high monetary rewards, that scientists do not walk out *en masse* from weapons laboratories. It is because they, like all other human beings in our time, are trapped in an obsolete structure of mankind, which our ancestors have bequeathed to us. They are part and parcel of a humanity divided into fractions, each of which enforces a certain moral code within it but acknowledges (whatever its leaders may proclaim) no such code for its relations with other nations. Scientists cannot hope to change dramatically this situation through passive non-co-operation of individuals, but they could—and, in my opinion, they should—contribute individually and collectively toward gradual reform of the world structure, ultimately to make possible a unified, harmonious humanity.

The political leaders of most nations understand that the arms race cannot last indefinitely and that, sooner or later, a form of international existence must be found which would permanently exclude war and thus make competitive armaments to win this war unnecessary. However, they postpone acting for such a new world system until after the victory of their economic or political ideology, which they confidently expect to win. Western nations believe—and point to historical experience in support of this belief—that popular control of governments by free elections is the best guarantee against military adventures into which personal or ideological dictatorships are almost inevitably drawn. The Communist leadership—also quoting historical evidence—believes even more ardently that the capitalist economy, with its recurrent conflicts between unrestrictedly competing economic groups, inevitably leads to war and that only world-wide acceptance of an economic system not motivated by profit can put an end to wars. Neither side is swayed by examples—some of them quite recent—which obviously contradict their generalizations. Both play for time in the hope that the other side bears seeds of inner instability and will sooner or later collapse. But mankind cannot afford to wait. Each additional year of the arms race means an additional chance of nuclear catastrophe. However small these annual increments, the risks of subsequent years add up, and eventually they will become overwhelming.

As scientists, we should realize that, at the time when the ideological lines which now separate the world into apparently irreconcilable camps were first drawn, the capacity of technology, founded on consequent application of science, to create wealth (by developing new sources of energy, by utilizing new raw materials instead of the scarce and unevenly distributed traditional ones, and by increasing productivity of labor through mechanization and automation) was much less obvious than it is now. We now know that with a well-developed science and technology (and given a sensible rate of population increase) enough goods can be produced to supply the basic needs of everybody, whether the economic system is the best one imaginable or not. Socialists believe, and will keep on believing, that the most equitable distribution of the products of labor can be achieved under a fully planned socialist system and that the increase in productivity will be fastest if it is not stifled by the requirements of profit earning; and they will quote chapter and verse for their belief. However, if we at all aspire to objectivity, as scientists must, none of us will deny that an economic system based on individual initiative and profit incentive has demonstrated, in America and elsewhere, a capacity to produce enough goods to achieve a fair standard of living for all. In the same way, the belief of many Americans that because of the absence of profit motive, a socialist economy cannot achieve a high level of productivity has been revealed as demonstrably wrong by the industrial development of the Soviet Union (and of the socialized segments of industry in Great Britain and other countries). In other words, although scientists may —and many will—maintain the conviction that one economic system is not only fairer but also more efficient than the other, they will see the difference in quantitative and not in absolute terms.

An important area affected by these considerations—an area in which (I believe) scientists of all countries have an abiding stake—is the economic advancement of educationally and industrially underdeveloped peoples. For these nations, the problems of political freedom and of the best economic system are overshadowed by the immediate and urgent need somehow to pull their masses out of the desperation in which want, undernourishment, and ignorance keep them. They need help from more advanced nations through capital investment, through the spreading of education, and through the training of technical personnel in order to be able to win the critical race between the increase in their agricultural and industrial productivity and the growth of their populations.

Scientists of all countries must feel an obligation to help human progress win against the blind forces of nature, not only for humanitarian reasons, but also because they are (or should be) aware that, in our age, peace and prosperity have become indivisible, worldwide problems, and that their own, at the present time more fortunate, nations cannot remain secure and prosperous as long as large parts of the world are not on the way to sound economic growth and consequent stability. I believe that scientists of all countries should combine efforts in this field, irrespective of their political allegiances and economic beliefs. They should urge their governments to join in providing such educational and technical help, leaving to the recipient nations the choice of economic and political forms in which to mold their aspirations. That such co-operation is possible has been demonstrated by the International Atomic Energy Agency.

I think scientists should explore together the possibilities of other creative international programs in technical, economic, and scientific areas.

It is more difficult—but perhaps not impossible—to mitigate also the second ideological controversy which now divides the world: the controversy between belief in individual freedom and in centralized political, economic, and intellectual guidance by the state. We in the West believe that the greatest over-all human progress can be achieved if maximum scope is given to the thoughts and aspirations of the individual. Admittedly, we cannot fail to see that, in certain respects, society can be weakened or its progress slowed down by the clash and pull of contradictory ideas, desires, and impulses of different groups within it. However, in the last reckoning, we expect this weakness to be more than balanced by a greater wealth of new ideas, by the spur of competition, and by the greater enthusiasm with which people work when they are permitted to pursue their own ideas and aspirations. On the other side, the view prevails that more rapid progress can be achieved by restricting severely the play of contradictory ideas and forces and by making everyone work in a co-ordinated program outlined by the ideological and political leadership of the nation.

Historical examples can be quoted in support of both views. Could we, perhaps, as scientists approach this controversy also in a quantitative, relative way rather than as a matter of dogma? As scientists, we have a common experience—that, in science, free inquiry and untrammeled exploration by individuals are the ultimate sources of the most important progress. The greatest scientific discoveries have come through efforts of non-conformist individuals who have asked heretical questions and boldly doubted the validity of generally accepted conceptions—be it flatness of the earth, the necessity of continued application of force to keep a body moving, the universality of time, or the continuity of matter.

On the other hand, we have witnessed, in our time, impressive examples of the organized application of science under central direction and cannot gainsay that such efforts can produce the greatest practical results in the shortest possible time—be it an atomic bomb or a Sputnik. These epoch-making feats are, however, not the great scientific breakthroughs the public thinks them to be; rather, they are spectacular practical exploitations of scientific breakthroughs which have occurred earlier, and often unnoticed, in the quiet of fundamental laboratories or at the desks of mathematicians or theoretical physicists.

Generalizing our experience as scientists, we may perhaps agree that society needs the free striving of individuals as well as organized collective effort. We will disagree among ourselves as to how much scope should be given to these two forces to achieve the greatest progress of science and technology—not to speak of an attempt to extend the same approach to other areas of human endeavor, such as the economic or political advancement of mankind. What we may, perhaps, agree on, is that the relative scope to be given, respectively, to the free, creative individual and to the organized collective should be considered as an empirical, experimental problem, without dogmatic prejudice.

I believe that the responsibility of scientists in our time is to bring into human affairs a little more of such skeptical rationality, a little less prejudice, a greater respect for facts and figures, a more critical attitude toward theories and dogmas, a greater consciousness of the limitation of our knowledge, and a consequent tolerance for different ideas and a readiness to submit them to the test of the experiment. These are the attitudes on which the progress of science has been founded in the past, and on which it remains based now. For scientists, there should be no final truths, no forbidden areas of exploration, no words that are taboo, no prescribed or proscribed ideas. Their common enemies are stubborn, preconceived ideas, prejudiced and closed minds—forces whose triumph would mean the end of science.

It is this open-mindedness that makes fruitful the international gatherings of scientists, such as the "Atoms for Peace" meetings at Geneva and the numerous other, less glamorous scientific conferences. Many believe that this type of co-operation cannot extend beyond purely scientific or, at best, technical areas. Perhaps they are right, but the sequence of Pugwash meetings is dedicated to the hope that this is not the case. Our meetings suggest that, with the increased penetration of science into all areas of human life, an attempt to extend to new areas the approach that has permitted men of different creeds and political or economic attitudes to work successfully together in science is at least worth making. The fact that we are meeting for the third time and that those of us who were present at the first and second conferences left them with a certain elation and a desire to return next time is encouraging. Admittedly, we should not fool ourselves or others. We are still only on the periphery of mutual understanding; we talk very cautiously, trying to respect the difficulties which our colleagues from different nations face in meeting with us. We try to talk mostly about our common beliefs, not to argue out our differences. Ultimately, we will be justified in speaking about a real success of our endeavors only if we are able to explore together the whole situation of mankind, analyze both the things that divide and those that unite us, and find a common program of action, not in pretended ignorance, but in full consciousness of these differences. This cannot be achieved immediately. It will require time and patience; but to aim at less would soon mean to find our movement at a dead end.

The first work in which scientists can and should combine their efforts, the first area of our common responsibility, is the education of peoples of the world and of their leaders to the understanding of the fundamental facts and implications of science for world affairs. The world must become aware of the essential irreversibility of scientific progress, of its dynamic character, of the impossibility of forcing it into prescientific patterns of national and international life. The leadership that will in the future neglect to give to the sober facts of science priority over ideological concepts, as well as over established traditions or national passions, will do so not only at its own peril, but—unfortunately—also at the common peril of humanity.

A second common effort of scientists could be aimed at securing greater understanding of the importance not only of scientific facts, but also of scientific methods of solving problems, for the future fate of humanity. No scientist will be brash enough not to recognize the continuing and often decisive importance for the behavior of nations of irrational factors—of political and national animosities; established traditions; ideological, racial, and religious fanaticisms. Many people, skeptical of the scientists' intrusion into political areas, say that they are ignorant or wilfully neglectful of all these forces and that they naïvely presume that mankind can be promptly reorganized on a rational, scientific basis. This is not true; but, without underestimating the power of the irrational forces in shaping the relations between men and nations, scientists must, *first,* analyze these forces with as much objectivity and open-mindedness as they can possibly bring to bear on these matters and, *second,* try to find a reasonable compromise between these forces and the arguments of reason. In this way they can help mankind to avoid the grave dangers of the dawning scientific age and to utilize fully its bright promises.

If, as scientists, we attempt to analyze the world situation, prognosticate on its future, and search for the best way to improve it on a broader basis, including psychological, emotional, and traditional factors in addition to the scientific and technological ones, we will undoubtedly disagree violently among ourselves. Some will believe that they are in possession of fundamental principles and of a methodology which can give them final answers to most if not all of the crucial questions. Others will bring into their attitudes a much greater skepticism if not agnosticism; still others, a religious or ethical attitude, based on belief in revealed truth. The one thing that could bring us together and could entitle scientists to a certain degree of leadership would be the demonstration of greater humility in the face of the unknown, greater tolerance of each other's points of view, and greater respect for mutual difficulties of nations and societies than are commonly displayed in the political and ideological controversies. For us, all these disagreements and conflicts must be overshadowed by common knowledge of the great challenge which the progress of science and technology places before mankind, of the dilemma of nations either to subordinate their political, ideological, and national aspirations to a common interest or to perish together.

One of our Russian colleagues at Lac Beauport remarked after a few days of this meeting, "I thought at first that we were wasting time. But now I see this is a different kind of conference. We have been accustomed to coming together to call for a certain thing, or to proclaim a belief we have held in advance, and then to disband. This is a conference to which people come with doubts in their minds to search together for truth. People must be made to understand this." This, of course, means only that we come to discuss public affairs as scientists and try to approach them in the same spirit in which we approach our problems in pure or applied science. To the extent to which our scientific analysis does lead to definite conclusions on which competent experts can agree, we can authoritatively address ourselves to the public. Our conclusions concerning the destructive possibilities of atomic warfare, the radiation dangers, or the nature and possibilities of scientific progress belong in this category. On these matters, the clear and present responsibility of scientists is to educate the people and their leaders.

I believe, however, that this does not exhaust our responsibilities. Beyond spreading information on positively established scientific facts, it also behooves us, as scientists, to assist mankind in finding adequate answers to the many new problems of national and international life in the scientific age. This requires first of all that we study these problems in the spirit of scientific inquiry. It would be improper for scientists to try to advise others without having first appraised the situation on their own and acquired as much positive knowledge of the facts as possible. Our second responsibility, then, is to study the impact of science on the affairs of man and search for objectively adequate answers to this challenge, as fearlessly and open-mindedly as we possibly can, accepting no ready-made answers from the outside—in the same way that scientists would approach any other unexplored area of knowledge.

The continued Pugwash program should thus, I submit, have a twofold aim. One is to educate peoples and governments in things we know; the other, to expand our knowledge. In both directions, scientists from different backgrounds, with different political and national allegiances, should be able to work together. Beyond these two responsibilities of *education* and *investigation,* some would suggest that scientists' responsibilities include also *action*—both individual action along the lines described at the beginning of this paper (refusal to work on military research, active participation in technical assistance programs, and similar constructive projects) and collective, concerted action to stop the arms race and prevent the misuse of science for destructive aims.

I argued at the beginning of this paper against the belief that the crisis into which the discovery of atomic energy has brought mankind could be prevented by the decision of individual scientists not to work on the new weapons. I am equally doubtful that similar individual decisions could now put an end to the arms race. It would be something different if scientists from all countries could agree on organized, collective action. However, this could occur only in an open world (for which Niels Bohr has called so eloquently in his well-known appeal to the United Nations). This world does not exist now and is not likely to come into existence in the foreseeable future. In its absence, no common action of scientists throughout the world can be reasonably contemplated. However, the world changes before our eyes, more rapidly than ever in history. What looks impossible today may become a possibility in the next generation. In the meantime, perhaps, we should not be afraid of thinking—and even talking together—about these matters.

Creation of a Suitable Climate for Disarmament

Desire for, and Fear of, Disarmament

It is easy to say that one or the other, or both, sides in the present world contest do not *want* disarmament and to explain in this way the persistent lack of progress in disarmament negotiations. However, for any state, Communist or not, the arms race is an unwelcome burden, particularly since it has been recognized by both sides that no "victory," in the old-fashioned sense, and therefore no material or ideological advantage of any kind, can be derived from the actual use of modern weapons. In addition, there is a widely understood danger, arising from the almost instantaneous delivery of modern weapons and from their enormous destructive power, of a war unleashed by a technological accident or human failure. This and the dangers inherent in the impending spread of scientific weapons to new countries are uncontrovertible reasons for all nations genuinely to desire effective disarmament.

There are, of course, in nations whose economies are based on private initiative, certain groups—entrepreneurs, workers, engineers, and scientists—whose economic security has become dependent on a continuous flow of military orders from the government. Although not expecting profit from *war,* these groups do profit from the continuation of the *arms race.* Like many other groups in the United States, they exert an influence on the policies of the American government, as witnessed by certain speeches delivered at Los Angeles during the presidential campaign. I think, however, that Mr. Khrushchev was right when he said, during his visit to the United States in 1959, that American industry could convert to peaceful production—as it actually did at the end of the Second World War—without a major crisis. Therefore, interest in continued military production is not the strongest motive determining the direction in which the American business community influences national policies.

Much more powerful than economic interests in the continuation of the arms race is the widespread fear of disarmament, which is

Paper presented at the sixth COSWA-Pugwash Conference in Moscow, December, 1960.

shared by almost everybody in America, with the exception of small groups of dedicated pacifists. Why this fear, if disarmament is almost universally considered as something desirable? It is the consequence of distrust of the intentions of the other side in the world contest. The American people are largely convinced that the ultimate aim of the policy of the Communist states is to destroy their own society—a destruction which, they believe, will also mean the transfer of supreme authority in their national affairs to foreign capitals. It would not serve much purpose to discuss here whether this apprehension is justified or not or to attempt to apportion the responsibility for its strength and persistence. What is uncontrovertible is that this distrust exists. It explains why a statement, such as Premier Khrushchev's remark, "We will bury you," is widely interpreted in America as meaning "we will conquer and destroy you" rather than, as it was probably intended to mean, "we will pass you so far in economic power that you will have to acknowledge the defeat of your system."

This distrust of the aims of Soviet policy accounts for the extreme sensitivity of Western public opinion (and Western political leadership) in the question of objective verification of compliance with any disarmament agreement which may be concluded with the Soviet Union. A disarmament treaty which could plausibly be presented to the American public as leaving open the possibility of evasion would not muster, now or in the near future, the public and congressional support needed for its ratification; and if it were to be ratified, its operation would be endangered by recurring suspicions and accusations of secret violations. This is why the demonstration, by a group of scientists, of the possibility of concealing nuclear explosions in underground caverns has made it impossible for the American government to sign an agreement for the cessation of nuclear tests without taking this possibility into account—at least by providing for future research on this problem—even if plausible arguments could be advanced why the Soviet Union is unlikely to engage in large-scale clandestine underground testing. Making the West less apprehensive on this account is a most important part of the problem of "creating a world climate favorable to disarmament."

COMMON INTERESTS VERSUS NATIONAL INTERESTS

The distrust in the West of the ultimate aims of the Soviet Union is further enhanced by conflicts between the immediate political designs of the United States and the Soviet Union in various parts of the world—Eastern Europe, the Near East, Africa, Latin America, and elsewhere. The ideological conflict between two types of societies is thus overlayed by "old-fashioned" conflicts of interests between competing powers. How can trust be established between powers whose long-range objectives are to destroy each other and whose immediate interests clash in every corner of the world?

Dr. Leo Szilard has said that there is no such thing as trust between states, only between individuals. I do not agree with him. I believe that trust does exist between states to the extent to which their common interests are stronger than any divergent or contradictory political aims they may pursue. This is clearly the case between, say, the United States and Canada. The trust between nations waxes and wanes as the community of their aims grows or declines. The establishment of trust needed for the acceptance of a reasonably controlled disarmament treaty depends therefore on the achievement of a sufficient community of interests between the East and the West.

It is widely postulated that the overwhelming common interest of the West and the East lies in the avoidance of wars and in the shared fear that a continued arms race may lead to war; consequently, it is suggested that the establishment of the trust needed for disarmament can become an "autocatalytic" process—each step toward disarmament engendering increased trust and thus permitting the next step. This, if I am not mistaken, is the official view of the Soviet government, and it is shared by most of my colleagues actively working for disarmament.

I believe that this is an optimistic oversimplification. The desire to avoid war and to end the arms race are "negative" common interests. I do not believe that by themselves they are strong enough to counteract the effects of the conflict of ultimate ideological aims and the clash of power interests. I believe that, in order to create a "proper climate for disarmament," both sides must also establish other common aims—partly negative ones, such as the abatement of the ideological conflict and the achievement of at least temporary stabilization of the local power conflicts; and more important, "positive" ones, such as the progress of world health and nutrition, the management of natural resources in the interest of all mankind, the exploration of space, and so on. The approach to disarmament itself, in a world where international trust is minimal, should be guided not, as it has been in the past, by considerations of political and strategic desirability (which, in the context of the power conflict, will inevitably lead to disagreement) but by considerations of the technical ease of controls and the possibility of accepting less than 100 per cent reliable control mechanisms without risking destruction in an attack or exposure to overwhelming diplomatic pressure.

I will consider first the possibilities of increasing international trust by developments in areas other than disarmament itself and later "autocatalytic" mechanisms of disarmament.

FOUR AREAS IN WHICH COMMON OBJECTIVES COULD BE ESTABLISHED

I am convinced that changes in national attitudes and policies in several fields could be of great assistance in making disarmament feasible. I have discussed these areas—of which I see four—at the previous Pugwash conferences, and I beg leave to repeat and develop some of these considerations here. The four areas are (1) the shelving of all attempts to revise the existing political borders or social systems in Europe and elsewhere; (2) the taking of the so-called underdeveloped areas out of the power conflict between East and West; (3) the dampening of ideological warfare; and (4) the seeking out and pursuing of positive co-operative programs.

(1) The Shelving of Territorial Disputes. Territorial disputes have been, in the past, one of the main—if not the main—source of wars; it is enough to think of the role of Alsace and Lorraine in wars between France and Germany or of Poland in the wars between Russia and its western neighbors. The present map of the world contains many disputed areas; for example, the boundaries, if not the very existence, of the state of Israel are under dispute. I think it is possible, and lies in the interest of both camps, to put all such territorial revindications into a "deep freeze" and to stabilize all now existing political boundaries. In the present mood of humanity, it would be useless to argue that, in our time, territory is not the main basis of national prosperity, as it may have been in the past, and that nations should peacefully agree to this or that revision of their boundaries. The expulsion of Greece from Asia Minor apparently strengthened rather than weakened that nation; and so did the loss of many colonial possessions by their European masters. The truncated state of Western Germany has developed into the strongest and most prosperous part of Western Europe, despite the loss of provinces in the East and the West and of all overseas possessions. But these object lessons of history do not reduce the passionate attachment of all nations to "every square inch" of their national territory. Ceding even the smallest part of the national real estate is considered a disaster, and even the catastrophe of a nuclear war is often said to be preferable to the loss of the smallest island or of a few square miles of desert.

This uncompromising attitude toward any change in national boundaries may be a survival from the time when sovereign national states were first born; certainly, the concept was unfamiliar to earlier, nomadic civilizations and, for that matter, to monarchs who have looked at national territories as private estates that could be bought, sold, or exchanged (as in the Louisiana Purchase and the sale of Alaska to the United States). Such deals, however reasonable and advantageous to both parties, are unthinkable in our time of unreasoned nationalism. Under these conditions, any change of national frontiers, however strong the arguments in its favor, is not achievable by peaceful means. Since mankind cannot afford the calamity of a nuclear war on behalf of any contested territory, both sides in the present conflict should decide—and make it clear to the world—that they will not lend their support to *any* territorial claim, anywhere in the world. This should apply especially to the eastern border of Germany. I believe that the West committed an error in not realizing, at the time the armistice lines were drawn, that it would be practically impossible to postpone the drawing of final frontiers until a future peace conference, and that, in the meantime, it would not be able to prevent the effective take-over of eastern German territories by the Soviet Union and Poland. Short of another world war, the present boundaries of Germany cannot be changed in any foreseeable future; the longer we wait with the acknowledgment of this fact, the longer the Germans are allowed to forget that it was their own attempt to remake the map of Europe by force that has brought about these territorial losses, the more difficult it will be for them to accept the present status quo as final. The West keeps this problem open at its own peril. It is alarming to hear German leaders assert that Western allies "owe" to Germany, not only support for German unification, but also the re-establishment of Germany's "historical frontiers" in the East. Nobody in the West believes that we have such a moral obligation; but the longer we maintain an ambiguous attitude in this question, the greater grows the danger of everybody becoming involved in a war over these "unsettled" frontiers. General de Gaulle seems to be the only Western leader who dares to speak out in this matter.

I referred above to German expectation of Western assistance in the reunification of Germany. This is part of a broader and even more dangerous threat to peace—claims not for territorial changes but for changes in the political and economic systems prevailing in certain territories of the world. Such claims are much more difficult to "freeze" than national frontiers, but I see no other way to stabilize peace. The West believes that the present political and economic system has been established in many states of Eastern Europe, not by the desire of their populations, but by the will of minorities supported by the military power of the Soviet Union. In the same way, the Chinese People's Republic considers the regime established in Taiwan as expressing the will of a minority group which settled in Taiwan after its defeat on the continent and which maintains itself by American military support.

It is clear that none of the nations in positions of power in such controversies is willing to permit the test of a free vote in an area where it runs the risk of losing it, whether in Eastern Germany, Taiwan, or Kashmir. It is conceivable that at some future time the world powers may become more amenable to local readjustments, but for this to happen, their position in the arms race must cease to be their main concern, so that losing a certain territory or tolerating a political change in it will cease to be a dangerous loss of strategic advantage for one side and an accretion to the military posture of the other side.

Until this stage is reached, I believe that all powers should accept, as facts of life which cannot be changed by force, not only the present national boundaries but also the existing political and economic systems. As I mentioned in my paper at Lac Beauport, this cannot mean the denial of sympathy which both sides bear toward what they consider legitimate desires of peoples but only a clear decision by both sides not to interfere with the internal processes that go on in these areas and a conclusive demonstration of actual adherence to this decision.

(2) *The Taking of Underdeveloped Countries out of the Power Contest.* Stabilization of the existing boundaries and political systems in the underdeveloped nations outside Europe is much more difficult than the stabilization of the situation existing on the European continent, because of the obvious transitory character of the political and social systems prevailing in many of these countries. Even the political boundaries of many new countries are artificial creations of colonial powers and may shift before these states crystallize into viable political entities. New social and political upheavals, such as those that have occurred in Egypt or Iraq in recent years, are likely. They are made almost inevitable by the intransigent policies of some white minorities. Such upheavals create a danger of war between the East and the West (as witnessed by the Lebanon crisis, which followed the Iraqi revolution, or by the present situation in the Congo) as long as both sides continue to see in them primarily an accretion or diminution of their military power—an attitude which calls for "choosing sides": "If you are

for Lumumba, we are for Mobutu." This choosing of sides is not primarily the consequence of different ideological preferences but of the bipolar power structure of our world. If both sides could come to the conclusion that it is in their own interests to reduce the danger of war resulting from taking sides in local conflicts, they could agree (and adhere to the agreement) not to support any side in such local struggles by supplying it with arms and military technicians and, even more important, not to use technological assistance as a weapon in the power contest.

The Soviet Union has proposed, at different times, an embargo on the importation of weapons into the Middle East and into Africa. It seems to me that these proposals should be considered favorably by the West. Of course, mere declarations or the signing of pacts prohibiting unilateral military support cannot be accepted without proper inspection; but this inspection should be less difficult to establish than inspection for disarmament, since it would not involve operations within the territories of the major powers.

The proposal for pooling technical assistance through the United Nations was denounced by Mr. Khrushchev when it was made by President Eisenhower at the height of the Congo crisis. This rejection was predictable since the American proposal was made *ad hoc* in a situation in which the buildup of Soviet assistance to the Congo seemed to forecast an absorption of the Congo into the sphere of Soviet influence. No suggestion was made by the United States that "what is sauce for the goose is sauce for the gander"; in other words, that unilateral technical assistance should be replaced by UN-sponsored common assistance in other regions of the world as well, regions where the United States has established successful bilateral operations. I believe that both sides should seriously consider the desirability—from their own points of view—of a universal agreement on this matter. Such an agreement could go far toward preventing the internal instability of many underdeveloped nations from generating sparks which could engulf the world in a nuclear holocaust.

Without such an agreement, the technical assistance operations are likely to develop, and already are developing, into major operations in the Cold War. No doubt, this fruitful competition is preferable to the sterile arms race; and many Americans welcome it in the belief that American productive capacity can easily meet the challenge. The Communist leadership has an equally legitimate reason to believe that a centrally organized and ideologically directed Communist society can surpass the unorganized, profit-motivated capitalist society in a technical assistance contest. Despite these visions of victory on both sides, I believe there are serious reasons to consider the pooling of technical assistance in a common agency as more advantageous to all donors as well as to the recipients of help. It is patently inefficient to treat the technological development of an underdeveloped country, not as a co-ordinated program, but as a battleground for competitive programs. For example, it would be completely ridiculous to treat the building of a harbor by Western technical assistance and the building of ships to use this harbor by the East as moves in the Cold War—let us say, by building the berths too small for the ships. The same applies to the construction of an electric power station by one side and a streetcar system by the other side, the irrigation of a valley by one side and the building of roads which make this valley accessible by the other. Even the building of two steel plants in India, one by the West and the other by the East, are, in essence, not competitive efforts but two parts of a common program aimed at reinforcing the industrial basis of the Indian economy. To make these activities deliberately co-operative would be merely to give formal recognition to their intrinsic nature. It seems probable that despite the advantages which some underdeveloped nations derive from playing the one side against the other in securing maximum assistance, most, if not all of them, would prefer not to play this dangerous game but to receive technical assistance from a UN organization, avoiding the danger of becoming stooges of one or the other power bloc. Without a "neutralization" of technical assistance, it will be practically impossible to satisfy the first above—mentioned condition for the elimination of underdeveloped countries as potential sources of war—the non-identification of the competing internal forces in these countries with the one or the other of the big power blocs.

No doubt, the replacement of bilateral assistance agreements by a common assistance pool cannot occur suddenly in a single sweep. Both states can, however, make increasing use of the already available channels for economic assistance, such as the UN Special Fund. The essential point is that the growth of collective technical assistance should proceed faster than the growth of competition in bilateral assistance; even more than in disarmament, an "autocatalytic" acceleration should be possible in this field, once the initial drive has been given a sufficient impulse. If, however, both sides wait, hesitating before the plunge, the inertia of growing competition in technical assistance will soon put obstacles into the path of co-operation as formidable as those which the accumulated inertia of the arms race has now put in the path of disarmament.

(3) The Dampening of Ideological Warfare. Both sides in the

world conflict have a deep conviction that they stand for a better form of society. In the West, this conviction centers on political freedom and the rights of the individual; many, but not all, in the West, believe that this freedom cannot be realized without a large degree of freedom in economic life (although very few now favor an entirely unbridled interplay of economic forces, without any control by the government as representative of the interests of the people as a whole). In the East, the conviction centers on greater social justice and greater capacity for technical progress of an economic system from which the individual profit motive has been eliminated and which is guided entirely by the government as representative of the common interest. The widespread power conflict between the Soviet Union and the West is exacerbated by this ideological controversy. Every time the power conflict flares up, an outburst of mutual vilification follows. It is difficult to envisage a truly peaceful coexistence of the two systems without a decision to avoid such mutual abuse. One cannot, and should not, expect each side to abstain from praising its progress and criticizing the ideological foundations and practices of the other side; but what one could strive for is the elimination of extremes of abuse and unreasoned wholesale disparagement.

Scientists are perhaps more likely than representatives of other groups to see the common problems confronting human society in all its forms and the need of a common approach to many of them; whereas others see only the problems that divide mankind and that these problems are approached in radically different ways by different societies. In science and in scientific technology, and perhaps even in industrial and agricultural management, men face problems raised by nature, which are the same everywhere. These problems can be solved only in certain ways, suggested by their objective nature and little affected by differences in economical and political philosophies—such problems as the irrigation of deserts, the combating of diseases, the exploration of space, and the penetration into the secrets of atomic structure. In addition, scientists are accustomed to being skeptical in respect to theories—even scientific theories based on experimental facts!—and to keeping an open mind when confronted with alternative interpretations. All scientific theories must be subject to revision and refinement if they are to keep in step with the increasing multitude of data supplied by observation and experimentation. As the role of science becomes increasingly important in modern society, the need for this critical and open-minded approach spreads over increasingly wide areas of human existence.

(4) *Identification and Pursuit of Common Positive Programs.* The scientific research in most countries has been, throughout history, a co-operative effort, extending over the whole Western world, without regard to political or religous allegiance. In recent times, this effort has spread over the whole world. For the first time in history, a world-wide community of people exists that is engaged in a common effort aimed at enhancing man's understanding of nature and increasing his capacity to control it for the benefit of all. The fact that individual societies have been, and still are, directing these efforts within their borders toward specific advantages of their own only superficially disguises the essentially world-wide, co-operative character of the whole scientific enterprise. Even in the military sphere, a new discovery which at first benefits one country (such as the discovery of the atom bomb) becomes the property of all mankind after a historically insignificant lapse of time. This is more obviously true of advances in non-military technology—new industrial processes, new agricultural techniques, new chemical products. Finally, in basic science, world-wide co-operation is automatic and universally recognized. All governments have abandoned the attempts which were made immediately after the war to keep the knowledge of certain advances in basic science restricted to their countries. Declassification of the research work on thermonuclear power in 1958 called an end to such isolationism in pure science. In pure science, there is no "Soviet science," "American science," or "German science" but only a single science of mankind. Governments have bowed before this fact and now assist international communication in pure science through sponsorship of international conferences, such as the two "Atoms for Peace" conferences in Geneva and other, less widely known, but no less important, international meetings.

International co-operation in science has existed throughout modern history; but as long as scientific research was a minor and politically insignificant activity of a small group of peculiar individuals, their co-operation was of little significance in human affairs. Now, however, science is rapidly growing in the number of people engaged in it, in funds expended on it by different nations, and in its importance for the future of mankind. Expenditure for scientific research has become a significant item in national budgets. Although the official justification of this investment remains the furthering of national interests in competition with other nations, all these efforts, wherever undertaken, serve, in reality, a common aim and help the advancement of mankind as a whole. It can be said that, *de facto,* all advanced nations now spend a significant proportion of their means on, and permit a significant proportion of their population to be engaged in, the pursuit of common interests of humanity. What is still lacking is open recognition of this already existing state of affairs, the official sanction of a situation which has arisen from the nature of the scientific endeavor rather than from the decision of political leaders. Nations pretend to serve only themselves but in fact serve all mankind.

With the increasingly close relation between the progress of science and the advancement of agriculture and industry, which is an outstanding feature of our time, the universal and co-operative nature of the scientific endeavor is spreading wider. I have already mentioned in the preceding section the *de facto* co-operative nature of the, on the face of it, competitive national efforts of assistance in the industrialization of underdeveloped countries. Here, too, what is lacking is the recognition of the co-operative character imposed by the nature of the problem and defying the continued intent of the several nations to further only their own aims.

If searching for appropriate areas of co-operative effort and fostering common programs could become a conscious aim of the several nations, despite their discordant ideologies and antagonistic political aims, many such areas could be found, and the proportion of national effort involved in co-operative international efforts could gradually become commensurate with the proportion dedicated to the achievement of competitive advantage in military or other fields. Nations could then begin to feel that, in a significant part of their total activities, they work not against each other but together in the common interest of mankind.

The International Geophysical Year was a strikingly successful example of such co-operative efforts. Others, equally or more important, can be devised. Space exploration is one of them. From the point of view of a cosmic observer, competition in space exploration must appear an absurdity. At present, national ambitions and prestige are deeply involved in such questions as which nation will be the first to land a man on the moon or to circumnavigate Mars. We are so hotly engaged in this competition that we cannot see how preposterous it is. Just imagine the astronauts from the East—or from the West—landing on a cosmic body and informing its inhabitants that they came there to prevent the astronauts from another part of the same planet, Earth, from coming there first and making the inhabitants of the planet their allies in an internecine conflict between humans on earth! It is true that nothing is so absurd as not to make sense in the heat of an ideological war; a Soviet poet has published a poem in which he anticipates the time when

Soviet astronauts will "establish in the empty world space and within the glowing stars the validity of Marxist teachings and the truth of communism on earth." But can any scientist, however ideologically committed, look at space exploration in this light?

Can one think of anything that would give a greater boost to mutual trust between the several parts of mankind than a common enterprise of space exploration? Just imagine the world following with bated breath the flight of a space vehicle manned by scientists from the East and the West! And why not? Only recently, newspapers were carrying the story of a trek across Antarctica in which Russian explorers were joined by an American scientist!

I believe that one of the most important contributions the scientists of the world can make toward the establishment of a climate favorable for disarmament is to foster international co-operation, not only in science itself, but also in the application of science to practical problems facing all mankind—in medicine, in agriculture, in the management of natural resources, in space exploration, in technical assistance to underdeveloped nations, and in many other fields.

In an article in one of the recent issues of the Russian magazine, *Science and Life,* Mr. Golubev let his imagination roam in describing engineering projects of great scope, bringing benefit to all humanity, which could be brought about if "peace could reign forever": the damming of the Behring Straits, the draining of a large part of the Mediterranean, the conversion of the Sahara into a fertile land, and so on. At the end of this article, Mr. Golubev said that all these plans could become feasible if only disarmament could be achieved. My strong conviction, which I want to convey here, is that we cannot wait for the beginning of such constructive co-operation until after disarmament. I believe we should start on this path—or rather, accelerate our progress on it, since first steps have already been taken—without waiting for disarmament. Programs which need no more mutual trust than already exists between the scientific and technological communities of the world should not wait for programs in which adequate trust is yet to be established. Constructive co-operation is the most effective means of changing the climate of international relations and making it more favorable to disarmament.

We recognized the need and possibility of this co-operation in the Vienna Declaration of two years ago. Since then, the progress of the arms race has made disarmament increasingly difficult; and the extension of the Cold War to new areas of Africa and Latin America has dampened hopes for constructive co-operation between the West and the East. The discouraging developments of these two years should not detract us from persevering in our efforts. As scientists, we cannot lose sight of our particular view of the world, in which the objective facts of science and technology play a greater role than they do in the world view of many others. The permanent implications of these facts cannot be erased by ups and downs in the international situation. These facts call for all nations not merely to "coexist" but to work together toward common aims. If they do not, they will perish together.

Autocatalytic Disarmament

I would like to return now to the possibility of using disarmament itself as a mechanism for the generation of trust. Others more competent than I have presented papers dealing with programs of disarmament and the methods of its control. All I can contribute to this discussion is to emphasize that in the determination of the sequence of disarmament steps due regard must be given to the technical ease of control and the relative need for high reliability controls. We have learned from the two year-long negotiations on the cessation of nuclear weapons tests that even a superficially easy control problem will become exceedingly difficult if the need for 100 per cent secure controls looms too large in the minds of one side, and the danger which may arise to national security from free-roaming inspection appears too large to the other side. As scientists, we know that 100 per cent secure disarmament controls are technically impossible and that a major sovereign nation will always be capable of carrying out certain clandestine developments if it is willing to invest sufficiently large funds and technical manpower in this enterprise. It is out of the question to establish mechanisms for certain objective detection of *all* nuclear explosions, including "kiloton range" explosions carried out in big underground caverns. Therefore, an agreement in this field has to be based on arbitrarily set limits of the number and size of explosions one is willing to let go undetected in order to see this preliminary step toward disarmament consummated. These limits must be set, not by the non-realistic criterion of "only explosions of no military significance should have a chance of remaining undetected," but by the more realistic, but much more sophisticated, criterion of "only explosions which, even carried out with a certain frequency, will not add significantly enough to the military power of the side engaged in the clandestine testing to make it worthwhile for that side to run the risk of detection should have a considerable chance of remaining undetected." If *systematic* and *significant* violations of the agreement would run a considerable risk of discovery, the theoretical possibility of occasional evasion will not appear worth the investment of large sums and technical manpower; and it can be reasonably hoped that no clandestine explosions will be carried out at all.

In the case of nuclear weapons tests, reasonable hope for ultimate agreement rests not in the possibility of improving detection mechanisms enough to permit certain detection of *all* nuclear explosions but on the fact that the advantage which may accrue from a small number of clandestine tests (and the disadvantages which could arise from the stopping of tests) is not decisive enough to make the acceptance of a fairly efficient, but far from 100 per cent certain, inspection mechanism impossible. Even so, a considerable change in the present mood of American public opinion will be needed, since distrust of Soviet intentions makes it responsive to the superficially plausible slogan, "No disarmament without 100 per cent certain inspection." In any case, we must learn from the difficult progress of negotiations on a test ban that in all future approaches to disarmament we must begin with steps which are relatively easy to control and, even more important, which would be acceptable to the two sides with only moderately efficient controls. The order of disarmament steps resulting from the application of this criterion may not be the one most desirable for either side and may appear illogical from political or strategic points of view; but it would have the advantage of facing the least technical difficulties. In chemistry, we are familiar with reactions proceeding by circuitous steps because the activation energies of these steps are lower than those obstructing the direct reaction path. Something similar applies to the progress of mankind toward disarmament.

If one is asked what kind of weapons nations will be most unwilling to relinquish without complete certainty that the other side has also relinquished them, the answer is obvious. These would be nuclear weapons of the kind which make possible instantaneous destruction of a nation not possessing them and could provide a powerful instrument of diplomatic blackmail, possibly leading to surrender without a fight. Other weapons, as we have seen from the history of two world wars, can be decisive only in large quantities, which are impossible to fabricate and stockpile in secret and which could be built after the beginning of the war to turn the decision in favor of the nation that was originally deficient in them.

It follows that the first step in disarmament cannot consist in the abolition of all nuclear weapons (or all means of their delivery). This is contrary to the desires of the peoples of the world, since their anxiety is based primarily on the fear of nuclear war. We would like to breathe freely in the knowledge that we are not threatened every minute of our lives by sudden destruction! Yet, the weapons in which the threat of destruction resides are weapons whose absence is particularly difficult to verify. It is a matter of common knowledge that stockpiles of nuclear weapons could be concealed in a small basement, anywhere in the country, and could not be detected except by betrayal.

It has been argued in other papers of this conference that one could allow each nation a certain number of nuclear weapons (and of means for their delivery in retaliation for an attack) and that such an arrangement would permit not only the destruction of a large number of "excess" nuclear bombs but also a far-reaching abolition of other weapons, under reasonable controls, short of absolutely certain discovery of every violation. The existence of permitted residual nuclear weapons would provide a carapace under whose protection a new, essentially disarmed world could emerge like a butterfly from an ugly cocoon. I am not competent to discuss the technical possibilities and merits of this proposal. I can only say that it makes sense to me.

Revising the Order of Priority

If we want to begin moving in the direction of complete disarmament, all nations will have to revise the order of importance of their different aims. This applies to both the Soviet Union and the West. The Soviet Union will have to decide that it wants disarmament more than it wants the preservation of its military secrets. From the beginning of disarmament negotiations, the Soviet Union has seemed to put greater stock in the preservation of secrecy than in the possibility of finding out more about the military establishment of the West through disarmament inspection. It has been willing to conclude disarmament agreements with no controls spelled out in detail. Even after the Soviet Union has agreed in principle to the need of effective controls, the impression persists that this agreement is a concession to the West and not the fulfillment of the Soviet Union's own desires. Since there is no reason whatsoever to suppose that the Soviet Union has more trust in the political leaders of the West than the latter have in the Soviet leaders, the only logical conclusion is that the Soviet Union considers continued secrecy of its military installations a more important asset than the possibility of knowing more about the military preparedness of the West. Because of the openness of Western societies, the additional possibility of locating Western military installations and identifying their equipment, which may arise from disarmament inspection, would be of relatively little value to the Soviet Union.

However, the technological developments of our time—space flight, long-range and infrared photography, radar, and so on—point toward the increasing difficulty of maintaining the secrecy of national military deployment. I do not think that any nation (or any individual) will be quite happy about the growing technical capacity of their own or foreign governments to invade their privacy, and some international agreements could perhaps be reached to protect it. However, no honestly meant agreement of this kind could be expected to protect the right of nations to maintain the secrecy of their military preparations. A renunciation of certain technical means of "finding it out" can be the result of practical considerations, such as the fear of undesirable consequences of certain methods of information gathering (e.g., the U-2 flights), but they cannot reflect a genuine belief in the principle of "legitimate military secrecy." The secrecy of military installations cannot be protected by any agreements once nations possess technical means to penetrate it! In the face of this technical change, it seems to me that no nation should base its plans for future national security too strongly on the waning asset of secrecy. We have learned, in our time, that the development of science and technology can affect many long-established political and ideological concepts. This happened, for example, to the age-old concept that the application of military force can lead to the victory of one society (or one political and economic doctrine) over another. Its obsolescence has been recognized now by the major powers of the West and the East—we hope that this recognition extends to the Far East—but there is still considerable reluctance to recognize and accept all the practical consequences of this new state of affairs.

It seems to me that scientists of all countries should be more clearly aware than anyone else of the consequences of the technological revolution and must help their societies toward the understanding of these consequences and the adaptation of national policies to all the facts of the technological age. They have done it successfully in making the practical consequences of a nuclear war widely understood by the often reluctant governments and public opinion. In recent years, the Soviet leadership has shown considerable understanding of the fact that the success of the Soviet system must from now on be based on its achievements in peaceful competition with other types of societies. I hope that similar understanding will develop also in respect to the secrecy of military preparations and will lead to the recognition that, from the point of view of the national security of each nation, it is more important to convince your potential enemy that you are not preparing to destroy him by a sudden strike than to keep secret your capacity for such a strike.

The need for revising the order of priority is, however, not all on one side. The governments of the West, too, have to learn not only to desire disarmament but to desire it more strongly than certain other things. The first of these things is 100 per cent reliable inspection of disarmament agreements—a desire reasserted by Vice-President Richard Nixon in his campaign speeches. As stated before, sticking to this requirement will make disarmament practically impossible. The matter of disarmament controls is, and will always remain, not an all-or-nothing proposition but a problem of degree—of the degree of probability of successful evasions and the degree of importance of such evasions for the balance of military power.

Another revision of the customary order of values is needed. The nations and governments of the West (and perhaps this applies to the East, too!) must learn to want disarmament more than the maintenance (or the achievement) of military superiority (or at least equality) in all kinds of weapons and in all possible war situations. This was a valid consideration as long as chances to win a war could be considered more important than chances to avoid war. What we need is the reversal of this order of priority! From this point of view, both sides should be willing to forego weapons whose main purpose is to increase the capacity for a first strike rather than assure their ability to deliver a second (retaliatory) strike. In other words, both sides should consider the renunciation, by mutual agreement, of weapons serving the "antiforce strategy," such as more powerful and numerous thermonuclear bombs, antimissile missiles, and so on. The same applies to weapons which would be of importance only in a protracted war of the type of the two past world wars, e.g., mine-bearing, roving submarines, intended for attacks on shipping.

In this connection, one should also consider that illicit possession of weapons which should have been eliminated in accordance with international agreements will be of no use in the diplomatic stage of a power conflict or in the striving for a national "image" of great strength. You cannot threaten the enemy or awe the world with nuclear missiles which are not supposed to be there! The existence of such weapons can be revealed only through actual attack or a demonstration at the height of a crisis, when diplomatic bargaining has broken down. This is another unorthodox consideration, which should be taken into account by those who clamor for 100 per cent safe inspection of every disarmament step. Continuing this train of thought, the West would be wise to reconsider its attitude toward agreements not to use weapons of mass destruction unless first attacked by such weapons. It would be foolish to consider such agreements as guarantees that these weapons will not be used if war should actually break out—this will be decided by other considerations—but a power that has signed such an agreement would deprive itself of the possibility of using such weapons for open or implicit threats, and this would help to avoid extremes of international tension.

A SUMMARY

These, then, are the developments which I believe are needed to create a favorable climate for the success of disarmament negotiations:

1. *Agreement on stabilization of the existing political situation* of the world. If an explicit agreement to this effect is impossible, some progress could be obtained by factual adherence to such policies by both sides.
2. *Insulation of the underdeveloped part of the world* from the power conflict between East and West by an embargo on arms and by channeling a gradually increasing fraction of technical assistance through a common agency, preferably under the United Nations.
3. *Increased emphasis on scientific co-operation and its extension to various areas of applied science, medicine, and technology;* the initiation of large common projects in the exploration and management of the resources of the earth and travel into space.

Parallel with this effort in areas not directly related to disarmament, an "autocatalytic" acceleration of disarmament negotiations could be achieved by aiming first for disarmament agreements reducing or eliminating weapons having no immediately decisive potentialities, which would not require extremely tight and reliable controls, and leaving to the end the complete abolition of weapons guaranteeing a minimum capacity for deterrence against nuclear attack. Another requirement for successful progress in disarmament is a revision of the order of importance attached to disarmament, on the one hand, and to other desiderata of national security, on the other hand. This would include a more reasonable approach to the problems of technical controls on the side of the West, accepting reasonable risks instead of requiring absolute certainty, and the reduction of the value attached to military secrecy in the East.

The Whole above the Parts

Different social or intellectual groups have left their imprints on human society in different eras. From the priests and prophets of ancient Israel and the learned mandarins of China to the merchants of Victorian England and the industrial pioneers of America, these groups have waxed and waned. Without considering each period in history as dominated by a single class, and all literature, art, religion, and science of this period as serving the interests of this class, one must admit that, at certain times, different—not necessarily economic—groups have exerted a particularly strong influence on the political behavior, moral ideals, and aesthetic standards of nations. In our time, science affects the fate of mankind in a way previously unknown. Leaders and movements whose power and ways of thinking have originated in a different age continue to control national policies, but they are forced to follow—sometimes too slowly and too reluctantly—the new directions imposed on them by the scientific revolution.

The contemporary social and national revolutions fall into the historical pattern—new groups emerging on the scene, claiming their rights, and wresting power from the older dominant groups. The scientific revolution is different. It does not change the power relations within mankind but the very background of human existence, the scenery against which the drama of history will unfold.

History knows periods of flourishing art, soaring religion, and accomplished statescraft and periods of their decay. Certain heights reached in the past have never been reached again. Science, however, knows only growth. Its advance may be faster or slower, but it is always moving forward, never backward. Theoretical concepts used in science may change their importance, but they are only temporary scaffolds which can be dismantled or replaced by new ones while the edifice keeps growing: Newtonian mechanics has been replaced by the more general relativistic mechanics; the wave and the corpuscle theories of radiation have been supplanted by higher synthesis in the quantum theory of radiation. But all these changes are components of a general process of growth. Nothing

Paper presented at the tenth COSWA-Pugwash Conference in London in September, 1962, and later printed in the *Bulletin of the Atomic Scientists,* December, 1962.

perishes in science but is incorporated into a more general system, brought closer to truth.

Another unique aspect of science is its universality. It does not belong to any one nation, as did other achievements of civilization. It is the first common enterprise of mankind. Admittedly, science does supply tools of competition to nations or systems; and it may be made to serve the military or economic interests of one group of men. But these are only temporary aberrations which do not touch the true essence of science. Scientists all over the world, irrespective of the society to which they belong and the religion or ideology which they profess, are working together on a common task: to give man a better rational understanding of nature and thus greater mastery over its forces. Every scientist knows in his heart that other scientists working in the same field, whether in his own country or anywhere else, are servants of the same cause—his colleagues and not his competitors or enemies. Achievements of science, kept secret for a while or utilized by certain economic or national groups for their special benefit, soon become the property of all mankind as a whole. Nuclear energy was first developed during the Second World War for the military advantage of America; but fifteen years later it is well on the way to becoming a power in the hands of all mankind, to be used for its general advancement or for its mutual annihilation. Giant rockets pioneered by the Soviet Union are rapidly becoming the property of other technologically advanced nations, to be used for mankind's growing mastery of space and understanding of the universe or for spreading the ravages of a nuclear war over the whole globe.

National, political, and economic revolutions merely change the power relations on earth. Their protagonists may dream of bringing about the unification of mankind in a common political or economic system, but these dreams are soon betrayed. The liberal revolution of the West engendered the subjugation of peoples all over the world to colonial rule and thus prepared the ground for new revolutions now sweeping Asia, Africa, and South America. The Communist revolution hoped—and still hopes—to unite mankind under a single economic system which its protagonists consider just and progressive, but it has already placed many nations under a rule not of their own choosing.

Scientific revolution, however, truly unites mankind. It gives nations the power of utter mutual destruction if mankind chooses to remain divided into traditional factions, each pursuing its own interests and powers as the *summum bonum;* but it also gives mankind the power to create satisfactory life for all if it chooses unity and peace. Other revolutions were born of scarcity, of exploitation of man by man and of nation by nation. The world resources and the productivity of mankind were limited, and the prosperity and power of one group had to be based on the poverty and subjugation of others. Seeking a greater share of this limited wealth—domination over greater natural resources, command over a greater pool of manpower—was the driving force of history and war was the final argument in this search.

Science has now made wars suicidal and destroyed their rationality as tools of politics. At the same time, science is showing mankind that the wealth at its disposal is not limited; new wealth can be created by human ingenuity. The energy of slave labor, once the foundation of civilizations, whether autocratic like that of the Pharaohs or democratic like that of the city-states of Greece, was replaced, in the first industrial revolution, by the much greater power of steam and electricity. But steam and electricity need coal or oil for their generation; and coal and oil are abundant in some parts of the world and scarce in others.

However, even nations abundantly endowed with fossil fuels will exhaust them in the not too distant future. Nuclear technology is providing mankind with a new source of power—nuclear fission—with resources far greater than those of fossil fuel. It has one novel aspect: it can be used anywhere in the world where technology is sufficiently developed, because transportation of nuclear fuel, in contrast to coal or oil, costs practically nothing. Still, access to fission fuels, uranium and thorium, is not open to all. But sooner or later, and probably sooner than later, science will find a way to utilize the energy of nuclear fusion. The raw material for fusion is water, abundantly available to all nations on earth. The age of fusion will be an age of equal access of all nations to sources of energy, and energy is the main ingredient of material progress.

In the area of nuclear energy, scientists have been able to think ahead of their nations and their leaders and ultimately to influence the policies of their states so as to reduce somewhat the danger of war, even if its spectre is still not exorcised. Scientists have not stopped the nuclear arms race, but neither have they been entirely unsuccessful in impressing on their societies and governments the need to seek, as a realistic and urgent aim, full disarmament and permanent peace—something which, until the scientific revolution of our time, was considered only a beautiful dream for the remote future.

The American people have been accustomed to considering the United States as an essentially peaceful state, trying to stay out of clashes between traditionally warring European nations and yet being repeatedly drawn into these conflicts. Peaceful coexistence with capitalist countries was offered by Lenin and again by Stalin. The difference between the attitudes of both countries in the past and their attitudes now is one of priority and urgency. Preservation of peace and a co-operative coexistence that can create mutual trust have only recently emerged as matters of first priority in the national policies of both sides. The pursuit of national, economic, or ideological interests, to increase the power and influence of a state or religious or ideological camp, have been in the past the basic and permanent aims of the units into which mankind is divided. These separate and divergent interests, be they economic or ideological, still exist and always will exist, but their pursuit, from now on, must be subordinated to the pursuit of the common interests of mankind: preservation of peace, rational husbanding of national resources, and full utilization of the creative capacities of science for the common interest of mankind.

I am certain that most scientists in the world share this conviction, however strong their attachment and their loyalty to their own nation, their faith, or their belief in a political or economic system. But in order to be able to influence the policies of our nations and the thinking of our leaders toward accepting this priority of common interests, scientists must do more than they have done so far to clarify their own ideas and to establish the priorities of their own commitments, even if this may mean contradicting official attitudes. We must realize more clearly and deeply than we have in the past that the scientific revolution has put all societies—Communist, capitalist, or whatever else they may be—before the same dilemma: live in peace and work together or perish. Before this challenge, other aims, however much cherished, must be set back. This new order of priority is not temporary but will continue forever, because whatever science has done to mankind is going to remain with it forever. Man will never unlearn how to build nuclear bombs or transoceanic missiles; he can only learn more about still more violent nuclear weapons, still more virulent bacteria, and still more deadly poisons. But neither can man's knowledge of how to create new sources of energy and new material wealth be lost; it too can only grow with time.

Scientists in both of the camps on which the peace of the world now depends were the first to proclaim what was then considered heretical by military leaders in the Pentagon as well as by those in the Kremlin: that against atomic weapons there can be no adequate defense; that with the discovery of nuclear energy, wars must end forever. Fifteen years later, this has ceased to be heresy. What is still considered as heresy is the assertion that the most cherished principles to which nations are devoted—whether national sovereignty, political freedom, or a "progressive" economic system—are not the supreme good whose furthering at all costs, even by force of arms, calls for the complete and unquestioned devotion of men. I do not ask my colleagues from the countries of the West to give up their belief in individual freedoms or political rights, or their hope that these freedoms and rights will be in the future shared by all men on earth; neither do I call on my Soviet colleagues to abandon their loyalty to an economic system in which they see a key to the most effective development of science and technology for the benefit of the people, or their hope that in the future all nations will recognize this system as superior and join it. What all of us must recognize and struggle for in our societies is that even these supreme national interests, values, and loyalties must now be subordinated to a paramount common aim: making the scientific revolution a blessing to mankind and preventing it from ushering in the doom of civilization.

The scientific revolution has effectively united mankind to perish or prosper together. The realization of this fact is slow and reluctant everywhere. It is up to scientists to speed it up, even if it may mean being accused of cowardice, a "better red than dead" attitude, in the West, or of heresy in the East.

Three years ago, at Vienna, we took the step from recognizing the universal need for disarmament to acknowledging that disarmament can be achieved only in a world free of wars; nations are now following, however reluctantly, this second leg of a journey away from historical tradition. It is time to take the next step. We must acknowledge that mere absence of war cannot be a stable state; that true peace requires more than mere coexistence of basically antagonistic camps. It requires extensive co-operation of nations and social systems in the pursuit of universal common strivings of humanity.

Knowledge of the dangers of nuclear war is by now general in all nations. That general disarmament is desirable has become the official attitude of all governments. That effective disarmament can be based only on complete abolition of war is being increasingly recognized. Only a few continue to think of war as a permanent element in the intercourse between nations and hope that nations can again harness war instead of abolishing it; but these "realists" of yesterday are the dangerous dreamers of today, and their influence is waning.

Scientists must now assume leadership in establishing the primacy of the common interests of mankind over all that divides men into separate camps. Scientists must be pioneers in establishing the one world of man, because theirs is the first common enterprise of mankind and because this common enterprise has now become the most important content of history.

ADDENDUM TO PART III

The Vienna Declaration

1. Necessity to End Wars

We meet in Kitzbühel and in Vienna at a time when it has become evident that the development of nuclear weapons makes it possible for man to destroy civilization and, indeed, himself; the means of destruction are being made ever more efficient. The scientists attending our meetings have long been concerned with this development, and they are unanimous in the opinion that a full-scale nuclear war would be a world-wide catastrophe of unprecedented magnitude.

In our opinion defense against nuclear attack is very difficult. Unfounded faith in defensive measures may even contribute to an outbreak of war.

Although the nations may agree to eliminate nuclear weapons and other weapons of mass destruction from the arsenals of the world, the knowledge of how to produce such weapons can never be destroyed. They remain for all time a potential threat for mankind. In any future major war, each belligerent state will feel not only free but compelled to undertake immediate production of nuclear weapons, for no state, when at war, can be sure that such steps are not being taken by the enemy. We believe that, in such a situation, a major industrial power would require less than one year to begin accumulating atomic weapons. From then on, the only restraint against their employment in war would be agreements not to use them, which were concluded in times of peace. The decisive power of nuclear weapons, however, would make the temptation to use them almost irresistible, particularly to leaders who were facing defeat. It appears, therefore, that atomic weapons are likely to be employed in any future major war, with all their terrible consequences.

It is sometimes suggested that localized wars, with limited objectives, might still be fought without catastrophic consequences. History shows, however, that the risk of local conflicts growing into major wars is too great to be acceptable in the age of weapons of mass destruction. Mankind must therefore set itself the task of eliminating all wars, including local wars.

Statement of the conference on "The Dangers of the Atomic Age and What Scientists Can Do About Them" (third Pugwash conference), held at Kitzbühel, Austria, on September 14–19, and at Vienna, on September 20, 1958 [later reprinted in *Bulletin of the Atomic Scientists,* November, 1958].

2. REQUIREMENTS FOR ENDING THE ARMS RACE

The armaments race is the result of distrust between states; it also contributes to this distrust. Any step that mitigates the arms race and leads to even small reductions in armaments and armed forces, on an equitable basis and subject to necessary control, is therefore desirable. We welcome all steps in this direction and, in particular, the recent agreement in Geneva between representatives of East and West about the feasibility of detecting test explosions. As scientists, we take particular pleasure in the fact that this unanimous agreement, the first after a long series of unsuccessful international disarmament negotiations, was made possible by the mutual understanding and a common objective approach by scientists from different countries. We note with satisfaction that the governments of the United States, U.S.S.R., and United Kingdom have approved the statements and the conclusion contained in the report of the technical experts. This is a significant success; we most earnestly hope that this approval will soon be followed by an international agreement leading to the cessation of all nuclear weapons tests and an effective system of control. This would be a first step toward the relaxation of international tension and the end of the arms race.

It is generally agreed that any agreement on disarmament, and in particular nuclear disarmament, requires measures of control to protect every party from possible evasion. Through their technical competence, scientists are well aware that effective control will in some cases be relatively easy, while in others it will be very difficult. For example, the conference of experts in Geneva has agreed that the cessation of bomb tests could be monitored by a suitable network of detecting stations. On the other hand, it will be a technical problem of great difficulty to account fully for existing stocks of nuclear weapons and other means of mass destruction. An agreement to cease production of nuclear weapons presents a problem of intermediate technical difficulty between these two extreme examples.

We recognize that the accumulation of large stocks of nuclear weapons has made a completely reliable system of controls for far-reaching nuclear disarmament extremely difficult, perhaps impossible. For this disarmament to become possible, nations may have to depend, in addition to a practical degree of technical verification, on a combination of political agreements, successful international security arrangements, and experience of successful co-operation in various areas. Together, these can create the climate of mutual trust, which does not now exist, and an assurance that nations recognize the mutual political advantages of avoiding suspicion.

Recognizing the difficulties of the technological situation, scientists feel an obligation to impress on their peoples and on their governments the need for policies which will encourage international trust and reduce mutual apprehension. Mutual apprehensions cannot be reduced by assertions of good will; their reduction will require political adjustment and the establishment of active co-operation.

3. What War Would Mean

Our conclusions about the possible consequences of war have been supported by reports and papers submitted to our conference. These documents indicate that if, in a future war, a substantial proportion of the nuclear weapons already manufactured were delivered against urban targets, most centers of civilization in the belligerent countries would be totally destroyed and most of their populations killed. This would be true whether the bombs used derived most of their power from fusion reactions (so-called clean bombs) or principally from fission reactions (so-called dirty bombs). In addition to destroying major centers of population and industry, such bombs would also wreck the economy of the country attacked through the destruction of vital means of distribution and communication.

Major states have already accumulated large stocks of dirty nuclear weapons; it appears that they are continuing to do so. From a strictly military point of view, dirty bombs have advantages in some situations; this makes likely their use in a major war. The local fall-out resulting from extensive use of dirty bombs would cause the death of a large part of the population in the country attacked. Following their explosion in large numbers (each explosion equivalent to that of millions of tons of ordinary chemical explosive), radioactive fall-out would be distributed, not only over the territory to which they were delivered but, in varying intensity, over the rest of the earth's surface. Many millions of deaths would thus be produced not only in belligerent but also in non-belligerent countries by the acute effects of radiation.

There would be, further, substantial long-term radiation damage to human and other organisms everywhere from somatic effects, such as leukemia, bone cancer, and shortening of the lifespan, and from genetic damage affecting the hereditary traits transmitted to the progeny. Knowledge of human genetics is not yet sufficient to allow precise predictions of consequences likely to arise from the considerable increase in the rate of mutation which would ensue from unrestricted nuclear war. However, geneticists believe that they may well be serious for the future of a surviving world population.

It is sometimes suggested that in a future war, the use of nuclear weapons might be restricted to objectives such as military bases, troop concentrations, airfields, and other communication centers and that attacks on large centers of population could thus be avoided. Even tactical weapons now have a large radius of action; cities and towns are commonly closely associated with centers of supply and transportation. We, therefore, believe that even a "restricted" war would lead, despite attempted limitation of targets, to widespread devastation of the territory in which it took place and to the destruction of much of its population. Further, an agreement not to use cities for military purposes, entered into in order to justify their immunity from attack, is unlikely to be maintained to the end of a war, particularly by the losing side. The latter would also be strongly tempted to use nuclear bombs against the population centers of the enemy in the hope of breaking his will to continue the war.

4. HAZARDS OF BOMB TESTS

At our first conference it was agreed that while the biological hazards of bomb tests may be small compared with similar hazards to which mankind is exposed from other sources, hazards from tests exist and should receive close and continued study. Since then, an extensive investigation by the United Nations Scientific Committee on the Effects of Atomic Radiation has been carried out and its authoritative conclusions published. In this case, too, scientists from many different countries have been able to arrive at a unanimous agreement. Their conclusions confirm that the bomb tests produce a definite hazard and that they will claim a significant number of victims in present and following generations. Though the magnitude of the genetic damage appears to be relatively small compared with that produced by natural causes, the incidence of leukemia and bone cancer due to the radioactivity from test explosions may, in the estimate of the UN committee, add significantly to the natural incidence of these diseases. This conclusion depends on the assumption (not shared by all authorities in the field) that these effects can be produced even by the smallest amount of radiation. This uncertainty calls for extensive study and, in the meantime, for a prudent acceptance of the most pessimistic assumption. It lends emphasis to the generally agreed conclusion that all unnecessary exposure of mankind to radiation is undesirable and should be avoided.

It goes without saying that the biological damage from a war, in which many nuclear bombs would be used, would be incomparably larger than that from tests; the main immediate problem before mankind is thus the establishment of conditions that would eliminate war.

5. Science and International Co-operation

We believe that, as scientists, we have an important contribution to make toward establishing trust and co-operation among nations. Science is, by long tradition, an international undertaking. Scientists with different national allegiances easily find a common basis of understanding: they use the same concepts and the same methods; they work toward common intellectual goals despite differences in philosophical, economic, or political views. The rapidly growing importance of science in the affairs of mankind increases the importance of the community of understanding.

The ability of scientists all over the world to understand one another and to work together is an excellent instrument for bridging the gap between nations and for uniting them around common aims. We believe that working together in every field where international co-operation proves possible makes an important contribution toward establishing an appreciation of the community of nations. It can contribute to the development of the climate of mutual trust, which is necessary for the resolution of political conflicts between nations and which is an essential background to effective disarmament. We hope scientists everywhere will recognize their responsibility, to mankind and to their own nations, to contribute thought, time, and energy to the furthering of international co-operation.

Several international scientific undertakings have already had considerable success. We mention only the century-old, world-wide co-operation in weather science, the two International Polar Years, which preceded (by seventy-five and twenty-five years, respectively) the present International Geophysical Year, and the Atoms-for-Peace conferences. We earnestly hope that efforts will be made to initiate similar collaboration in other fields of study. Certainly, they will have the enthusiastic support of scientists all over the world.

We call for an increase in the unrestricted flow of scientific information among nations and for a wide exchange of scientists. We believe that nations which build their national security on secrecy of scientific developments sacrifice the interests of peace—and the progress of science—for temporary advantages.

It is our belief that science can best serve mankind if it is free from interference by any dogma imposed from the outside and if it exercises its right to question all postulates, including its own.

6. TECHNOLOGY IN THE SERVICE OF PEACE

In our time, pure and applied science have become increasingly interdependent. The achievements of fundamental, experimental, and theoretical science are more and more rapidly transformed into new technological developments. This accelerated trend is manifest alike in the creation of weapons of increased destructiveness and in the development of means for the increased wealth and well-being of mankind. We believe that the traditions of mutual understanding and of international co-operation, which have long existed in fundamental science, can and should be extended to many fields of technology. The International Atomic Energy Agency, for example, aims not merely at co-operation for establishing facts about atomic energy but also at helping the nations of the world to develop a new source of energy as a basis for the improvement of their material welfare. We believe that international co-operation in this and other fields, such as economic development and the promotion of health, should be greatly strengthened.

The extremely low level of living in the industrially underdeveloped countries of the world is and will remain a source of international tension. We see an urgent need to forward studies and programs for the effective industrialization of these countries. Not only would this improve the level of living of the majority of the population of the world; it would also help reduce the sources of conflict between the highly industrialized powers. Such studies would offer fruitful scope for co-operative efforts between scientists of all nations.

The great increase in the ease and speed of communications and our increasing understanding of how the forces of nature influence the living conditions of nations in different parts of the world show us, in a way not previously possible, the extent to which the prosperity of individual nations is connected with, and dependent upon, that of mankind as a whole and how rapidly it could be increased by common international effort. We believe that through such common effort the coexistence between nations of different social and economic structures can become not merely peaceful and competitive but to an increasing degree co-operative and therefore more stable.

As scientists, we are deeply aware of the great change in the condition of mankind which has been brought about by the modern development and application of science. Given peace, mankind stands at the beginning of a great scientific age. Science can provide mankind with an ever increasing understanding of the forces of nature and the means of harnessing them. This will bring about a great increase in the well-being, health, and prosperity of all men.

7. The Responsibility of Scientists

We believe it to be a responsibility of scientists in all countries to contribute to the education of the people by spreading among them a wide understanding of the dangers and potentialities offered by the unprecedented growth of science. We appeal to our colleagues everywhere to contribute to this effort, both through enlightenment of adult populations and through education of the coming generations. In particular, education should stress improvement of all forms of human relations and should eliminate any glorification of war and violence.

Scientists are, because of their special knowledge, well equipped for early awareness of the danger and the promise arising from scientific discoveries. Hence, they have a special competence and a special responsibility in relation to the most pressing problems of our times.

In the present conditions of distrust between nations and of the race for military supremacy which arises from it, all branches of science—physics, chemistry, biology, psychology—have become increasingly involved in military developments. In the eyes of the people of many countries, science has become associated with the development of weapons. Scientists are either admired for their contribution to national security or damned for having brought mankind into jeopardy by their invention of weapons of mass destruction. The increasing material support which science now enjoys in many countries is mainly due to its importance, direct or indirect, to the military strength of the nation and to its degree of success in the arms race. This diverts science from its true purpose, which is to increase human knowledge and to promote man's mastery over the forces of nature for the benefit of all. We deplore the conditions which lead to this situation, and appeal to all peoples and their governments to establish conditions of lasting and stable peace.

September 19, 1958

PART IV
HERETICAL THOUGHTS

Science and Education for Peace

In the past, nations have been governed by soldiers (such as Caesar), philosophers (such as Marcus Aurelius), churchmen (such as Richelieu), historians (such as Thiers), or even artists (such as Paderewski), but never by men trained in natural science. Benjamin Franklin is the rare example of a man of scientific background who has acquired at least a certain prominence in his country's political affairs.

The Need for Scientifically Trained Personnel in Public Life

It has rapidly become commonplace to state that we live in a scientific age. To ensure the proper national and international administration and utilization of the vast powers which science has given to man and to prevent the destruction of our civilization by misuse of these powers are the foremost political problems of our time. That is not to suggest that science should replace law as the normal educational background of our statesmen and politicians but to state that it would be for the good of this country—and the same applies to other countries as well—if legislative bodies and public officialdom included at least a sprinkling of people thoroughly trained in pure and applied science.

When the questions of atomic energy came up for discussion in Congress, scientists were alarmed at the thought that no senator or representative could be expected to possess a knowledge of the underlying scientific facts—such as the distinction between nuclear and electronic phenomena—and therefore be trusted with correct appreciation of the military and political implications of atomic power. "Primers" in nuclear physics were hurriedly prepared and distributed among congressmen; but how many legislators have used them?

On one occasion, when the international control of atomic power came up for discussion in the Senate Committee on Atomic Energy, two senators disagreed violently, one obviously overestimating the importance of "denaturation" of fissionable materials as a method of making atomic power plants "safe," the other calling this procedure a "hoax." The question is by no means academic—an intelligent

From the *Pi Lambda Theta Journal,* May, 1947.

approach to it is essential to a wise decision as to whether large atomic power plants can be left in private or national hands without danger to peace. A modicum of understanding of the scientific facts (such as familiarity with the concept of isotopes) would have provided the two senators with a common background on which to base their judgment. In the United Nations Atomic Energy Commission, almost the only progress so far has been achieved in the Scientific and Technical Subcommittee, in which the presence of physicists and technicians, taking part in the discussions together with diplomats, has rapidly provided a considerable area of understanding as to the basic facts of the situation. After the severe jolt which the cause of international control of atomic energy suffered in the open discussion before the political forum of the Security Council, it was rapidly referred back to this subcommittee. If scientists and engineers were fully entrusted with this all-important problem, they would perhaps not promptly produce a universally acceptable solution; but at least much misunderstanding and shadowboxing would be eliminated. No deadlocks would have arisen over such dummy issues as the "dismantling" of existing atomic bombs—as if they could not be reassembled on short notice! The legal background of our statesmen and diplomats and their lack of scientific training naturally cause them to overestimate the importance of such legal questions as the use of the veto in the punishment of violators at the expense of such questions of substance as whether atomic power plants are important, whether they can be kept under control by inspection, and whether international management is necessary to prevent such plants from becoming a menace to peace.

To the assertion that scientists and engineers (assuming the latter do not represent private interests!) have a common ground and could agree on controversial political and international problems more easily than politicians, the answer is often given that when scientists (or other experts) are asked to give their opinions they are often found to disagree violently among themselves. This criticism—which was often read in the newspapers in connection with the political activities of American scientists in the last two years—is based on a misunderstanding of the scope and methods of science. It takes a man with a certain understanding of science to formulate problems in such a way that a scientist can answer them. Laymen—for whom a scientist is a kind of a minor prophet—usually ask questions which scientists are not able to answer *as scientists* but, for profit motives or because of a desire to be helpful, often try to answer anyhow.

An Educational Task

To make possible an intelligent utilization of science and technology for the public good and permit the full use of the advantages of scientific methods in arriving at national and international decisions, what is needed is not only that some scientists be transformed into politicians and public officials but, most of all, that politicians and public officials (and public-minded citizens in general) acquire a certain understanding of the basic principles, working methods, and possibilities of science. To provide basic general education in science is an important task of our educational system. At present, science is studied mostly as preparation for certain technical professions. Those who go into commercial, legal, or political fields soon drop from their minds as unnecessary ballast the science they learned at school. Of the remainder, the majority consider their scientific training merely as a set of tools given to them to solve practical problems which they may encounter in life and thus for advancement in their professions. Only a few, even among those who become academic scientists, are imbued with what we may call the "spirit of science," the belief in scientific principles and methods as the basis for their general attitude in social and political life. Many a good scientific worker becomes completely irrational and unscientific when asked to give judgment outside his narrow field of specialization.

The teaching of science in our schools should be directed not only toward the communication of facts and skills which might prove useful to future chemists, doctors, or metallurgists, but toward implanting in future generations an understanding of the basic principles of science, of the usefulness of the scientific method in solving problems, and of the basic human values inherent in this method which make it applicable to situations outside narrow technical fields. We need to educate not a generation of scientists but a generation of Americans fit to live in the "scientific age."

Science versus Technology

The principal obstacles preventing science from taking the place it should occupy in the minds of Americans are the confusion of science with technology and of discovery with invention—the widespread misconception that Edison was a great American scientist, that science consists in building new gadgets or inventing new processes for the attainment of such definite practical aims as flying faster, curing tuberculosis, or making bigger and better bombs.

Science in its purest form, in which it has the most fundamental and universal importance for man, is the quest for knowledge for the sake of knowledge. Without being blind to possible practical applications, the true scientist does not let them interfere with the direction of his quest. This quest is a probing in the dark and therefore essentially aimless; whenever the scientist works with a definite aim in mind—such as making an atomic bomb—he is an engineer and not a scientist. A combination of humility in the face of the endlessness of the unknown into which he is probing with a readiness to venture into this unknown and the consequent realization of the temporary and relative character of all present-day knowledge, concepts, and beliefs are the basic ingredients of the scientific attitude. Expanded into social and political fields, the same attitude should make for a refusal to believe that any institutions or political theories are sacrosanct and unchangeable and should promote open-mindedness and tolerance.

One Science in One World

To nature there is no difference between nations and creeds. A scientific problem, whether explored by a white man, a Japanese, or a Negro, will yield the same results. Communists have tried to establish a difference between bourgeois and proletarian science; the Nazis, one between Jewish and Aryan science. Both attempts were ridiculous failures. There is only one science, as there is only one nature. The internationalism of science is not a historical accident. Secrecy and suspicion (generated by war and now extended into peacetime) can disrupt the international co-operation of scientists but cannot "nationalize" science. Even "compartmentalized," to use the ugly military term, scientists all over the world remain members of one team, working toward the same aims. A man in Japan, Germany, or Russia working in the same field as I, photosynthesis by green plants, is not my enemy, not even my competitor, but my co-worker. Every amount of progress he makes is my progress; every new fact he discovers helps his colleagues in America as much as it does his friends in his own country. This denial of national exclusiveness, this feeling of unity and spirit of co-operation of all nations, is another ingredient of the true scientific attitude which the teaching of science should help to implant in students.

Respect for Facts and Readiness for Judgment

A scientist can get nowhere, in the long run, by twisting or ignoring facts, by making statements which are not confirmed by tests, or by insisting that he is right after the tests made by others do not support him. Respect for facts and readiness to be judged by others and to accept their judgment are essential elements in the makeup of a scientist. No force and no propaganda can make right out of wrong in science. A scientist must be able to criticize himself severely, because with each publication he exposes himself to criticism by others, a criticism which no friendship, national solidarity, or economic interest can mitigate. This is the attitude which scientists should—and the best of them do—also bring into the discussion of social and political problems. The teaching of science in schools and colleges could implant this attitude not only in all scientists but also in future civic leaders and statesmen. Such a changed attitude is needed to avoid the tragic paradox of civilized mankind's being unable to use the achievements of science for its own benefit, because of a refusal to renounce obsolete political and social concepts even in the face of facts which clearly prove them to have lost their former rationality and usefulness.

To conclude, with the advent of the atomic age, two new tasks have arisen for our educational system. In the first place, it must provide for an increased dissemination of scientific *facts,* and prepare a greater number of scientifically competent men and women to be scattered through different trades and professions, even those where scientific training was unheard of in the past. In the second place, and vastly more important, education must communicate to the new generation some elements of the true scientific spirit—the feeling of an essential unity and solidarity of all mankind, tolerance for other peoples' opinions, respect for facts, and readiness to submit oneself to criticism and to be judged by the test of unprejudicial experiment rather than to seek success by force and craft and acclaim by propaganda.

The danger of a material and spiritual collapse of our Greco-Judaeo-Christian, urban, technological civilization is acute. We cannot expect the present leaders of nations to create anything better than a temporary patchwork of peace. Even at that they may fail, and the next generation may find itself, before it has had time to grow up, one third dead, one third crippled, and one third destitute, roaming over the ruins which were once our cities and the lifeless, radioactive plains which were once our fields. But if reasonable and skillful leadership succeeds in avoiding the catastrophe of another war in this generation, it still will be the task of generations to come to convert the precarious, temporary peace structure into a permanent, stable edifice of a world community of law. The preparation for shouldering this responsibility seems to me the foremost task of education in our time.

23

The Labors of Sisyphus

According to Greek legend, the titan Sisyphus was condemned by the gods—for crimes forgotten by most who remember his cruel punishment—to eternal frustration. He had to roll a heavy rock up a steep mountain; whenever the top was in sight, the rock slipped and rolled down the slope, and the weary titan had to start from the bottom again. The task of getting Congress to understand the position of science in national affairs proves to be as frustrating as the labors of Sisyphus.

Since 1945, scientists and their friends in Washington have attempted again and again to impress upon Congress the simple truth that all technological progress (including that of military technology) is but a by-product of the advancement of fundamental knowledge, that one cannot foster the first and neglect the second, any more than one can produce good fruit or flowers without growing plants with healthy roots and leaves. To let basic scientific research deteriorate because it does not promise immediate practical return would be as improvident for a nation as to neglect planting the seeds for next year's harvest (even if he might still be cashing in on last year's crop) would be for a farmer.

Occasionally, in the past five years, it looked as if this simple truth had been grasped even by the most obstinately anti-intellectual senators and congressmen. It seemed that scientists could go back to their laboratories in the certainty that the moment they turned their backs on Washington Congress would not declare science a luxury. And yet, in August, 1951, the Appropriations Committee of the House of Representatives did just that. Because—as the committee stated in its report—fundamental research does not promise immediate contributions to the national mobilization effort, it clipped no less than 98 per cent from the appropriation for the National Science Foundation! Just enough money was left to pay the salaries of the head of the Foundation and a few of the secretaries. Having thus relieved the nation's budget of a "luxury" item of fourteen million dollars, the Appropriations Committee can now go back with vigor to signing multibillion-dollar blank checks for "practical" science, such as new atomic plants and "mystery weapons."

As this is written, there is still hope that this crippling decision

From *Bulletin of the Atomic Scientists,* October, 1951.

may be reversed in the Senate and that at least a part of the National Science Foundation funds can be restored in the final version of the appropriation bill. But the mere fact that in the year 1951 a prominent group of American lawmakers could decide that basic scientific research was a luxury, and could wait for national support until better times, is a frightening experience for all believers in the collective wisdom of a self-governing nation.

Since the end of the war, the vastly enlarged scientific research establishment of America has been kept going, in universities and research institutes, largely through the financial aid of the armed forces and the Atomic Energy Commission. Military leaders possessed the enlightened understanding of the importance of pure science which Congress lacked. With their billion-dollar budgets, they were able to take in their stride the spending of many millions every year on the support of fundamental research. However, the Office of Naval Research (and its counterparts in other branches of the defense establishment) have never considered themselves as more than temporary "angels" of American science, ready to relinquish this role with the appointment of a legal guardian.

Presently, with full mobilization at home and actual war in Asia, the military have less and less time, energy, and even money to devote to playing foster mother to American science. When, after many vicissitudes, the National Science Foundation Bill was finally passed by Congress and signed by the President on May 10, 1950, military agencies were all set to transfer most of their activities to the new Foundation and to make available to the latter the experience they had accumulated. They must now feel like a policeman who, upon having assisted the birth of a baby in the police cruiser, is refused admittance at the doors of a hospital—while his radio calls for all hands to help in pursuit of armed robbers.

The strength of America is based on mass education and wide diffusion of scientific and technical knowledge. The American military establishment is superior to that of other countries, not in the toughness of its soldiers or the discipline and morale of its combat units, but in their better developed intelligence, more extensive technical training, and modern equipment. Therefore, the American military establishment has a vital interest in a high level of general and technical education, as well as in vigorous pure and applied scientific research in this country. In an emergency, the military establishment may therefore feel that it is within its competence to assist national education as well as the national research effort. It is, however, as abnormal for the armed forces to finance scientific research outside their own laboratories (such as the Naval Research Laboratory or the Aberdeen Proving Grounds) as it would be for the military to take over continuous financial responsibility for American colleges other than West Point and Annapolis!

Imagine what would have happened if private funds and civilian authorities had failed to provide means for higher education in the country and the Army and the Navy had jumped in with grants from their vast budgets! This would have caused loud protest across the nation. People would be horrified to see their colleges treated as if they were institutions required primarily for the success of universal military training. Yet, a similar situation has existed for six years in American science! It has been treated as if it were deserving of national support only to the extent the military think they need it—and the military were therefore considered the proper agent to support it. For six years, scientists have worked to convince Congress that this attitude is fundamentally wrong. They had finally succeeded when the National Science Foundation was established—but it seems that they may have to start from the bottom again.

Perhaps, actions such as that of the Appropriations Committee reveal not merely an indifference to science but a subconscious hostility to it. This hostility has existed ever since science became an important agent of progress and change in society; it has not diminished but rather increased in times when science has produced its most spectacular achievements. An antiscientific, antirational state of mind was a part of the wave that carried Hitler upward; not for nothing was his accession to power celebrated by the burning of books! An antiscientific trend is clearly discernible in the present policy of the Soviet dictatorship, completing its path from pretended extreme rationalism to utterly irrational belief in verbal shibboleths and the divine wisdom of anointed leaders. It is well to remember that even in America, whose rise to prominence has been manifestly tied up with successful practical application of science to all branches of human activity, the distrust of the scientific mind and disbelief in the general soundness of scientific methods is widespread. Here, too, science is a magic bird whose golden eggs everybody wants to enjoy but whose free flight into regions inaccessible to most makes it a suspect creature. Yet it is in these flights into the unknown that science gathers the food that makes it grow. Put it into the most luxuriously equipped henhouse, and after a while there will be no more golden eggs.

A Speech for the President

Introduction

The following article was written in November, 1953, when an "Operation Candor" was much discussed—and finally dropped—by the Eisenhower administration. The only outcome of this discussion was the President's "Atoms for Peace" speech to the United Nations on December 8, 1953, which ultimately led to the creation of the International Atomic Energy Agency (IAEA) in Vienna.

America has passed in its short but eventful history through many critical hours. At its birth one hundred eighty years ago, the founders of the Republic engaged in a war of independence, armed with little more than courage and faith. Some eighty years later, their descendants fought a civil war for the preservation of the Union in the face of doubts and discouragements that seemed grave indeed at the time. Thirteen years ago, we were attacked by a coalition of two mighty aggressors, and practically the whole of our navy was put out of action on the first day of the war. Yet, in all these crises, the President could tell the nation that if it but persevered for a few months or years in its exertions and sacrifices victory would be won, and won it was. It falls to me to be the first President to tell our nation that, though at peace, we are confronted with a peril graver than any America has faced in past wars and that it is not in our power to remove this threat by exertions that have proved adequate in the past. It falls to me to call for efforts and sacrifice without hope of reward, except at the end of a long and dark journey of unforseeable duration.

When the perils we and the whole world are facing began to take shape before my eyes, I became convinced that the American people must share in the understanding of their danger and in the search for salvation. I consulted with many advisers. Some of them warned me against talking to the people as long as I could not show them an easily accessible path to overcome the perils. They told me that to reveal the threat and to confess that I had no easy prescription for its removal would be bad politics. What weighed more with me was the warning that people, in their alarm and anxiety, might turn

From *Bulletin of the Atomic Scientists,* January, 1954.

to leaders who would promise them cheap salvation—demagogues who would tell them that all we need to regain security is to renounce the freedoms on which this Republic has been founded, root out subversion, silence the opposition, and find safety in "fortress America," no matter what happens to the rest of the world.

These warnings, I had to agree, were based on a shrewd assessment of human reactions to danger and frustration; but I could not listen to them. To do so would be to betray democracy and to endanger our chances of ultimate salvation. This Republic is based on the belief in the wisdom of the people. We hold that, properly informed, people will act with celerity, reason, and imagination. To renounce this belief would be to choose enlightened dictatorship or oligarchy in preference to democracy.

I was further told that in order to inform the American people properly about the perils of the situation I would need to disclose technical and military information which would benefit our enemies. This argument, too, I had to reject. The facts which I need to state are simple and undoubtedly known to our adversaries. The all-important fact is that several nations now possess weapons of immense power, capable of destroying Western civilization which is largely embodied in cities, harbors, and industries. We have these weapons; the Soviet Union has them; and in due course, all nations willing to invest the considerable, but not prohibitive, funds and manpower needed for this purpose, will own them. We, who started the development of atomic weapons with a two-billion-dollar wartime investment, have acquired a head start of several years. In consequence, we now possess more atomic bombs than anybody else; in fact, we have enough atomic bombs of the original, so-called fission type, each capable of destroying a city of two hundred thousand people, to wipe out every such city on the face of the earth. We have successfully tested a land-bound device and will soon test an air-deliverable bomb of a new type, hundreds of times more destructive than the bombs we dropped on Japan in 1945—the so-called hydrogen, or "fusion," bomb, capable of wiping out a city of a million people. Presently we will be able to deliver these engines of destruction to any point on earth by long-range aircraft; we are engaged in the development of other means of delivery less subject to interception, such as guided missiles and long-range rockets, and expect that this research will be successful within a few years.

According to unmistakable evidence, based on objective monitoring devices and other intelligence data, the Soviet Union possesses both fission and fusion weapons with a power similar to ours. We do not know their number but have no reason to assume that it will not be, in a few years, sufficient to visit universal destruction on the cities of the world. The time when the Soviet Union first came into the possession of atomic weapons undoubtedly was advanced by the penetration of Soviet agents into our wartime plants, but to assume that more effective secrecy measures could have given us then, or can give us now, lasting security is a delusion against which I must warn the American people. We must assume that the Soviet Union—and other nations—will be able to match any advances we may still make in the development of more destructive weapons and of more perfect means for their delivery. Science, in its world-wide progress, has opened in our time the gates to a new world—the world of atomic nuclei—for all mankind to enter. This discovery has given the leaders of all nations access to weapons capable of destroying, at a moment's notice, all that our civilization has built up in centuries. There is no way to avoid the danger arising from these discoveries, either from the United States or from any other part of the world, short of putting an irrevocable end to war.

We are now engaged in a contest with a totalitarian dictatorship having at its disposal the resources of one-third of Europe and one-half of Asia. This dictatorship, presently armed with atomic and hydrogen weapons, is committed to a political philosophy according to which the rest of the world is its preordained and implacable enemy. Its leaders hold that the struggle of the two economic systems they call "socialist" and "capitalist" is the central theme of history in our time; that this struggle may be interrupted here and there by a truce; but that it will not end until one of the two contestants is wiped out. They believe that the capitalist countries represent a decaying and doomed way of life, and are bound to plot against the socialist world, but that in the end they will be destroyed and communism will rule the world. We can only hope that the Soviet leaders will soon realize that this interpretation of history—if it ever had any validity, which we doubt—has now become patently wrong and that in the face of the threat of universal destruction the common interest in survival should take precedence over the divergent aspirations of conflicting political and economic systems. We will welcome any sign, direct or indirect, that their minds are not closed on these matters; but as long as no such change is visible, it is not in our power to terminate the contest which never was of our choosing. Although we are ready to seize on every opportunity for the relaxation of world tensions and willing to seek agreement with the Soviet Union in any area in which accommodation appears possible, we cannot allow those efforts to interfere with our efforts to maintain and increase the capacity of the non-Communist world to deter and resist the drive of Soviet imperialism.

The danger of war is threefold. A temptation for an aggressor to unleash it lies in the defenselessness of our teeming cities and rich industries, in the ferment of the awakening nations of Asia and Africa, discontented with their poverty and resentful of the West, and in the disunity of the nations of the West. We must deal patiently and resolutely with all three dangers to strengthen the stability of the divided world.

To combat the first danger, we must proceed with the utmost dispatch with the development of means for the early detection, and effective interception, of any attempt to paralyze our country by a sudden atomic attack. All the inventiveness of our scientists and engineers must be mobilized for this purpose. This program calls for considerable increase in our defense expenditure, but it would be irresponsible gambling not to undertake it. The success of our program will not guarantee us complete immunity, but it will make the aggressor doubt whether an attack would be decisive—and hold his hand. With the same purpose in mind, we must accelerate the lagging program of dispersing the most vital industries and making American communication lines at home and abroad less vulnerable. This, too, will require large expenditure and—what is worse—distasteful but necessary restrictions of economic and individual freedom.

The second danger calls for the greatest efforts on our side to help the new nations of the East improve their standard of life and find hope for the future in a political framework which, even if it does not always satisfy our ideal of democracy, at least leaves the way open for a development toward this ideal. An auspicious beginning was made with the so-called Point Four program; I intend to call for a substantial increase in this program. To be successful, it must be pursued on a humanitarian basis, without any military or political conditions attached to it. The new nations are proud of their recently acquired independence and are sensitive to anything that reminds them of the past political domination by the West. We must assure them, by word and deed, that we respect their independence and their status as equals in the family of nations and that they are free to work out in their own way the solution of their internal and external political problems. We must stand by to help them with our wealth and experience in the hope that no nation will choose of its own volition the road to tyranny when hope burns brightly ahead on the path of freedom.

Stalin has bequeathed to his successors a blueprint for success based on what I have called the third danger we have to combat—

the disunity of the West. The experience of recent wars, from the First World War to the Korean War, has demonstrated that the greatest danger of war lies in an aggressor's belief that the opposing nations will not stand together, particularly if the initial object of the attack is a weak and remote country—Serbia, Poland, Finland, or Korea. In the age of atomic weapons, when standing up to aggression means risking the destruction of your own homeland, this calculation may acquire a new plausibility for an aggressor. Only by establishing a perfect unity of the free countries of the Western world can we put an end to such dangerous hopes. To advance toward this aim would require a continuous and increasingly intimate involvement of American military forces in Europe and generous economic co-operation with the Western nations. From the American point of view, these commitments often appear as one-sided assistance in reconstruction and defense, extended to other nations at the cost of American wealth and manpower. In the eyes of many Europeans, still suffering from the losses and destruction inflicted upon them by the last war, the same program often appears as their involvement in a power conflict between two great empires over which they have no control—an involvement carrying with it a certainty of total destruction in the case of war and providing no guarantee that they will not be abandoned to their fate whenever America decides to withdraw from Europe. We must remove, once and for all, at least this last argument for European neutralism. We must make it plain by acts that we stand with these nations in an organic, permanent, and indissoluble union—one for all, all for one. The difficulties, now seemingly unsurmountable, of the integration of France, Germany, and England into the West European Commonwealth, will disappear as soon as this American commitment is established beyond doubt.

These three policies will place an increasing burden on our shoulders and will not give us a full guarantee against war. However, we can hope that they will provide the maximum attainable deterrent to an outbreak of war. Beyond this, we have to place our hope for the future on developments within the Communist-dominated area, developments which we cannot influence except indirectly, by presenting to the world a picture of unity, steady progress, and readiness for conciliation. History or mass psychology cannot tell us how long a political and economic system based on fanatic adherence to a dogma utterly at variance with the realities of life can survive and whether such a system can, with time, become reconciled with the outside world without a last violent attempt to assert its supremacy. Yet, our safety from the disaster of an atomic war ultimately depends on the answer to these questions, which all the tremendous economic and military power of the United States and its allies cannot decisively change or resolve. This is what I had in mind when I said that we will need patience, wisdom, and imagination to remain steady and resourceful through long years of insecurity and frustration.

Concentration on the danger to peace from aggressive and dogmatic Soviet imperialism, however justified it is at present (or may remain for many years to come), should not make us forget that atomic weapons have raised the specter of the destruction of civilization in *any* full-scale war that may break out from now on into all future time.

Western civilization is about two thousand years old, if we count from the spread of Christianity, a few hundred years older if we count from its beginnings in Greece and Judea. It has established its home from the Pacific coast of Siberia to the capes of Brittany and Spain that jut into the Atlantic and from Alaska to the Tierra del Fuego; it has spread its influence far beyond this area, over all of Asia and Africa. Its scientific and technological achievements have been unprecedented in the history of the human race; its accomplishments in art, philosophy, and religion equal, if not surpass, those of any other of the known world civilizations. We who are a part of it are accustomed to thinking of it as *the* civilization—the final emergence of mankind into a road of universal and continuous progress, after several promising but local and ultimately unsuccessful attempts. Yet, we can now clearly see that this civilization, too, can perish and that it carries the seed of its destruction in itself. Other civilizations have been destroyed by the sword of invaders; ours is headed toward suicide in an atomic war. However, the true reason for the downfall lies—perhaps always—in inner inadequacy, in a lack of capacity for moral growth to match economic and technical progress. The danger that threatens our civilization is that three thousand years after the Sixth Commandment and two thousand years after the Sermon on the Mount, we still have not learned how to eliminate murder from our society. We—and that includes all nations of Christian tradition—have made murder a crime within the state but have permitted it to remain a virtue when carried out on a large scale between states. War has been with mankind since its beginning. Scorned by the wise and abhorred by the compassionate, it has been glorified in legend, art, and history as the highest test of human virtue and loyalty. It has been accepted throughout history, and is still accepted by most people, as an inevitable calamity in the life of nations.

This is the point at which our generation is faced with the choice of making a clean break with the past or facing a catastrophe. It is true that war could be justified in the past as a means of furthering national aspirations; but this justification does not exist now. The court at the Hague, the League of Nations, and the United Nations have been attempts to build a world without wars. These attempts have proved unsuccessful—the United Nations itself became a belligerent in Korea! For this failure, equal responsibility must be accepted by all nations. Admittedly, militaristic and totalitarian regimes were the culprits who broke the pledges in 1914, 1939, 1941, and 1950. Nevertheless, we, as well as the other democratic nations, bear full responsibility for having permitted the lawlessness in the world to continue into the atomic age. Like all other nations, we have felt more strongly about the advantages of full national independence and sovereignty and the diminution of the freedom of action which would arise from our submission to an international order than about the necessity to put an effective end to war. We are accustomed to hearing ourselves praised for our peaceful policies, and feel justified in blaming other nations which have had the misfortune of falling into the bondage of totalitarian dictatorship for being a threat to peace; but from a long-range point of view, we should be concerned with our own errors, and ask what we ourselves can do to prevent the present Soviet-Western war threat from being only the first in a series of threats, each facing mankind with an increasing likelihood of complete annihilation. We must turn our backs decisively on the era of nations that are a law unto themselves and seek the lodestar of our political thought and action in the establishment of a world system of law.

Therefore, I propose that the American people solemnly dedicate themselves to the establishment of permanent peace on earth, in their own enlightened interest and in the fulfillment of their duty to mankind. As a first step toward this aim, I propose that Congress proclaim in joint session the conviction of the American people that man's acquisition of power over the atom has made all resort to war criminal madness, certain to bring down the whole of our civilization. We should call on all nations to join in this proclamation. To establish the sincerity of this proclamation and our intent to implement it, I propose that we consider an amendment of our Constitution, making the establishment of a world order of law and the achievement of the abolition of war the aims of our Republic and providing a constitutional way for America to become part of such an order. Legal provisions of this kind have already been introduced in the constitutions of other countries, the most recent being the Netherlands. Finally, as a first step toward an approach to this idea, possible in the present divided world, I propose that we enter into consultation with like-minded nations about steps which could be taken without delay toward initiating such an international order.

The peril is great. If we meet its challenge and remain wise, patient, and humane, we can hope to emerge at the end as saviors, not only of American lives and American freedoms, but of the civilization we share with many other nations.

What Is a Security Risk?

The decision of the Personnel Security Board of the Atomic Energy Commission on the J. Robert Oppenheimer clearance case seems, on the face of it, absurd and, in its implications, dangerous. The absurdity consists in asserting, without reservation, that Oppenheimer is a loyal citizen and has handled secret information discreetly in the past—and then describing him as a "security risk" not fit to receive secret information in the future.

The direction in which Oppenheimer's political sympathies, his awareness of security dangers, and his personal associations have developed since 1943 deprive this conclusion of logical conviction. The Board argues that the security criteria have changed since 1943; but what must be measured in this case is not the 1954 clearance regulations against the ones used in 1943 but ten years of actual performance in highly sensitive and responsible positions against the purpose of all such regulations—the safeguarding of the nation's secrets. The concept of "security risk" was substituted in the Presidential directive of April 27, 1953, for that of "disloyalty" as the criterion in government clearance procedures. It has been generally assumed that this was done with a clear and legitimate purpose: to be able to withhold persons from government employment who, while not disloyal, could conceivably become dangerous sources of information leaks because of personal characteristics (addiction to alcohol or narcotics, abnormal sexual habits, tendency to loose talk) or because of circumstances beyond their control (such as having close relatives in Communist-ruled countries).

In this context, the term "security risk" obviously cannot be applied to Dr. Oppenheimer. He is not a drunkard, dope addict, or pervert; he has no close relatives behind the Iron Curtain. As to any inclination toward loose talk, the Security Board has explicitly affirmed his exceptional discretion in handling secret information. The only way in which the term "security risk" can be used in the Oppenheimer case is by widening its meaning far beyond its original scope—from persons who may compromise secret information to persons who may advocate a "wrong" kind of political or military decision. This new definition of "security risk" would include all those whose moral, religious, or political convictions, alleged lack

From *Bulletin of the Atomic Scientists,* June, 1954.

of candor, or other character traits might cause the government to distrust them; in particular, those whom it might consider capable of tendering advice, for any reason whatsoever, weakening the military and political posture of the United States in the world. A mere "lack of enthusiasm" for certain military policies—even before their official adoption—can be taken as prima facie evidence of such a "security risk." Here is where the decision of the Security Board ceases to be absurd and becomes dangerous.

It is the right of every government to select its advisers and to exclude from their number anybody it distrusts. The public can criticize the wisdom of the administration in taking Mr. A. into its confidence and refusing to listen to Mr. B.—in the same way that it could criticize the dismissal of General MacArthur as commander-in-chief in the Far East and his replacement by General Ridgway; but nobody can deny the administration the right to choose its advisers as well as its military commanders. If the present administration, or the present leadership of the AEC, do not want to take Dr. Oppenheimer into their confidence, despite his uncontested professional qualifications and unique experience, for any or all of the reasons set forth in the majority opinion of the Security Board, this is their privilege. This does not mean, however, that the government has the right to brand him a "security risk."

A large proportion of government servants are protected by civil service regulations which make it impossible to separate them from their job except for cause. Justly, this protection does not cover those responsible for policy decisions. By the scope and importance of advice which top scientific leaders are now called upon to provide, they are placed in a similar position. The Board itself acknowledges that it is often impossible for such scientific advisers of the government to restrict themselves to purely technical matters.

Those who are fully familiar with new scientific developments and capable of extrapolating these developments into the future cannot but think—and feel strongly—about the political, military, and moral implications of scientific and technical progress. Sometimes they are the only ones able to pose certain pertinent questions of our national policy; more often they see such problems in a light different from that in which they appear to others. Their opinions and predictions are not voices of absolute wisdom; they can go wrong even in technical prognostication, not to speak of the projection of their thought into politics or human relations. Nevertheless, the newly gained importance of science in national and international affairs calls for their advice as a significant contribution to the planning of national policies. When the national government, on whom the responsibility for final decisions rests, decides to act contrary to the opinion of a group of scientific advisers (as it did in the case of the H-bomb development), it can do so only by using other advisers with the same scientific background and competence. In such a situation, the government may lose its trust in the judgment of some of its scientific aides and decide to dispense with their services. It is, however, in the highest national interest that, at every moment, the government should find the most highly qualified scientists in the country ready to give them the benefit of their thinking, frankly and without fear that an erroneous or simply unpopular judgment may bring them, at some future time, into the predicament which the proceedings of the Security Board have brought upon Oppenheimer.

It would be a grave new departure in American political life if the established method of replacing unpopular or dissenting members of the administration team should become that of branding them "security risks." A secretary of the treasury may argue for the reduction of military expenditures because of a strong belief in the importance of a balanced budget. If he is overruled, he may decide to resign, or the President may even ask for his resignation. What would we think, however, if to obtain this resignation the secretary should have to be adjudged a "security risk"? Would the general esteem in which the secretary may be held by his colleagues, and the influence he may wield among the nation's business community, be considered aggravating circumstances, since they might lead to a delay in the adoption of the increased military budget?

Not so long ago, a large part of American public opinion reacted strongly—and its reaction was endorsed by the President—against the assertion of Senator McCarthy and his political allies that the administrations preceding the present one represented "twenty years of treason." It would be a dangerous deterioration in American political life if every change in policy were to be accompanied by proclamations that the previous policymakers were not merely wrong but disloyal, or at least, security risks!

The Board says in its report that it asked the AEC whether it would not be adequate to let Dr. Oppenheimer's consultantship expire (as it was due to on June 30) rather than institute proceedings for the revocation of clearance. Two objections seem to have been made to this sensible proposal. First, it was said that "without the positive act of withdrawal of access he (Dr. Oppenheimer) would continue to receive classified reports of atomic energy activities even though his services were not specifically and currently engaged." This statement is in flat and unexplained contradiction to Mr.

Lloyd K. Garrison's view that Dr. Oppenheimer's clearance would automatically lapse with his contract on June 30. Even if, as seems natural in the case of a scientist of Oppenheimer's eminence, whose loyalty and discretion have been upheld, the clearance were to be maintained by a specific decision after the end of the contract, this would not give him access to any secret information except that which the AEC itself would want to communicate to him. Furthermore, it must be remembered that the decision of the Board is nowhere motivated by a possibility that Dr. Oppenheimer might compromise secret information but only by the apprehension that his decision or advice might be wrong or unreliable.

The second argument against letting Oppenheimer's consultantship lapse without explicit revocation of clearance is that other government agencies or AEC contractors, who are accustomed to following the AEC's lead in these matters, would be likely to use Oppenheimer's advice on secret matters. Here, too, the only logically possible wrong is not that of compromising information but of giving wrong "advice" on policy questions.

Is it conceivable that a decision of the AEC not to use his advice on such matters, if taken with the approval of the national administration, would remain unknown to other policymaking agencies, such as the National Security Council or the Defense Department? If, on the other hand, the distrust of Oppenheimer's judgment is a decision of the AEC alone, is it not presumptuous of the AEC to impose this distrust on all other government agencies (and incidentally to make it difficult for any future Commission to use his services) by substituting for an expression of their personal appraisal of his value as adviser a quasi-judicial finding?

It seems strange that the Security Board has considered the two unconvincing arguments sufficient to exclude the alternative (which they themselves said they "would prefer to report")—to advise the AEC that they found no reason to classify Dr. Oppenheimer as a "security risk" but that, if the AEC finds that it cannot trust his judgment, it should simply not renew his consultantship. The report says that this alternative would have been possible only if the Board members had been "allowed to exercise mature practical judgment without the rigid circumscription of regulations and criteria established [for them]." If this statement is correct, it constitutes the strongest possible criticism of the existing security regulations. It suggests that these regulations make a decision called for by "mature practical judgment" impossible and enforce a decision which puts an unjustifiable blot on the reputation of an outstanding American scientist, generates distrust between American scientists and their government, and endangers the readiness of government advisers—scientific and non-scientific—to give the government the benefit of their best thinking. This decision must confuse and dismay our friends and comfort our enemies abroad. If such a decision is rendered inevitable by the security rules, then there is something fundamentally wrong with these rules, and rapid revision is called for, before the indefinite extension of the concept of "security risk," introduced in the majority decision, is permitted to become an established principle in American political life.

26

Science and the Humanities in Education

Wide—and loud—discussion of the proper roles to be given in our high-school and college education to the scientific and humanistic subjects has been under way since last fall when Russian prowess in science and technology was demonstrated by the launching of the first two earth satellites. The ensuing campaign for radical changes in our school curriculums has two aspects. In one battle, science and the humanities appear as allies in an attempt to reduce the emphasis put by "progressive" education on so-called life-adjustment studies and shift it back to where it rested earlier in America, and still rests in Europe (including Russia!), namely, on the study of scientific and humanistic subjects, such as mathematics, physics, chemistry, history, geography, languages. The other battle is between the sciences and humanities (meaning by "humanities" all social and humanistic disciplines) for allocation of the time and attention left over by the life-adjustment studies. Neither of these two contests, as such, is the main subject of this paper. Rather, I should like to emphasize the necessity of *integration* of all three areas of education—life-adjustment, science, and the humanities, all three of which, I believe, are needed to produce generations of Americans fit to survive and capable of leadership in the scientific age—a new era in history into which we were rushed headlong with the discovery of atomic energy thirteen years ago.

"Life-Adjustment" in the Atomic Age

The idea that education should not merely provide familiarity with the subject matter of science and the humanities—biology, physics, chemistry, mathematics, history, literature, and so on—but also produce useful members of a free society is a valuable contribution of American thought to modern educational theory and should by no means be discarded because we have become frightened by the success of Soviet education in producing large numbers of competent scientists and engineers. It has justly been pointed out, in defense of education for citizenship, that intensive study of science

An address delivered at the Forty-fourth Annual Meeting of the American Association of University Professors at Denver, Colorado, April 25, 1958 (printed in the *Bulletin* of the Association, Vol. XLIV [Summer, 1958]).

and the humanities did not prevent German youth from following in droves the ascientific ideology and the inhumane policies of an Adolf Hitler. Upon leaving the Gymnasium, with its rigid time-filling curriculums in mathematics, history, science, and languages, German youths found themselves helpless and bewildered in facing the realities of life; the unaccustomed democratic forms of political and social life seemed to them to provide no solutions for the problems posed by the loss of the First World War and later by a worldwide economic crisis and unemployment. Rationality and political sense, which should have been acquired from the study of science and history, appeared of no use in dealing with these harsh facts of life.

Whatever may be wrong with the present life-adjustment education in America, the fault is not in its aim. It is not that the school has no responsibility for producing men and women able to live responsibly and happily in a free society but that the kind of society to which our schools have been trying to adjust our children is not the society which they will actually face. The society which our education presupposes is gone with the wind. As the genteel society of the aristocratic South was blown away by the storm of the Civil War, so our present supposedly stable and permanently prosperous American society is being torn to pieces by the gale of a world-wide triple revolution—social, racial, and scientific.

Despite the violent concussions of two world wars and of the great depression, the preceding generation of Americans lived under essentially constant or only slowly changing conditions. Since the reforms of the New Deal, the aim of our society—which at times seemed close at hand—has become idealized in the "Great American Dream," the dream of a permanently and universally prosperous nation, slowly but steadily growing in wealth, in fairness to all its members, and in equitable distribution of the superabundant products of a highly productive industry and agriculture. The remaining evils of poverty and racial discrimination seemed destined to slow disappearance. This was to be an example of a successful society for all the world to admire and imitate. Our unique economic power and wealth were to permit us to assist occasionally in straightening out the tangled affairs of the rest of the world and to help to rehabilitate the victims of hunger and disease in less fortunate countries. (In one, more recent, version, our superior military might was to keep a *pax Americana* in the world.) Our boys and girls had to be educated to fit into this fundamentally sound and stable society. We were on a high plateau, not at the beginning of an arduous climb. We needed no crusaders, no fighters, no great dissenters. What we needed were good community and family men, conscious of their responsibility in local and national affairs but also of their natural right to security and happiness, men willing to help in removing the remaining minor evils of our society but essentially satisfied with it.

If this dream did not run directly into conflict with reality, much could be said in favor of our present system of education. One could have argued that in order to sustain, and gradually improve, the American society it is not important that schools graduate large numbers of restless intellectuals dissatisfied with the existing state of affairs or a large number of scientists and engineers who could rapidly advance our technology. Schools were not there to stimulate the growth of exceptional minds, to produce great reformers and great scientists, but rather to spread widely the moderate amounts of knowledge that prepare people for living a satisfied and useful life, working effectively within the existing framework of economic and social relations and enjoying the leisure which technical progress is placing at their disposal.

Unfortunately, it was not fated that parents who had caught a glimpse of this enviable state should bequeath it to their children. Instead, the world, and an America inextricably involved in it, has entered an era of instability, of breath-taking changes, of terrifying dangers and blinding promises—a world about which only one thing can be said for certain: that it will look different ten years from now from the way it looks today, and different twenty years from now from the way it will look in ten years. For any nation, even the most powerful one, to hope that it can quietly enjoy political stability and economic prosperity in the midst of the vast turmoil which now engulfs the earth would be a dangerous delusion. We are educating a generation whose main aim in life will be security, in an age which is, and will remain, abundant in everything but security. In the words of Reinhold Niebuhr, we will have "to live for decades in a continuation of responsibility, anxiety, tension, and frustration, an experience our young nation has never before faced." [1]

Several violent developments have contributed to make our time perhaps the most dynamic era mankind has known. First, there is the great social upheaval of our time, which began with the Russian revolution forty years ago—rebellion against the old titled and moneyed ruling classes (and, as Djilas has emphasized, the growth of new ones). Then there are the even greater racial upheavals, which have been slowly gathering speed since the beginning of the century (when Japan first emerged as a world power by defeating a great European power, Russia) but which the nations of the West immensely

[1] Niebuhr, "A Predicament We Share with Russia," *New Leader*, XLI (1958), 10.

accelerated by their two internecine world wars. Third, and perhaps most important, there is the great scientific revolution, which began a century or two ago but has been tremendously accelerated in the last twenty-five years. This revolution has provided man, for the first time, with quantitative and qualitative understanding of his status on earth and also with the means of changing this status beyond recognition, either by endangering the very viability of the human race through the spread of radioactivity in a nuclear war or by assuring slow but certain progress toward worldwide prosperity in a clearly foreseeable future.

In the midst of this violently changing world, it is more than ever necessary that our sons and daughters receive an education that will make them fit to face life, to encounter its realities with understanding and hope, to make them willing and capable of bearing the hardships which may be needed to fashion a brighter future for the world and to steer mankind away from the path of destruction upon which it is now launched. This reality is not the one to which progressive education has been addressing itself in the past years and to which it still tries to cling. We need not abandon "education for life," but we desperately need to adjust it to rapidly changing ways of life. The next generation of Americans must be capable of taking active part in the creation of new political, social, economic, and technological forms of human existence and not merely of fitting more or less passively into the existing ones. They must combine an abiding faith in the basic values of our society—freedom, equality, justice—with imagination, refusing to regard the existing relations, the accepted rules and the prevailing habits of American (and international) life, as permanent and justified just because they were born into them. The ship of humanity is passing through a violent storm on an uncharted sea. Education for life must be an education in how to stay on deck and steer the ship through narrow channels between threatening rocks into a safe harbor, not how to ride out the storm in a comfortable stateroom behind battened hatches.

It is not easy for our youth, growing up in the midst of an economy of abundance, in an essentially successful society, to recognize the need for struggle, hardship, and privation in striving for a better future. Russian or Asian youth grow up amidst poverty and squalor, and for them, the dream of a better future for their own people and for the world has a direct, irresistible appeal. They are taught that the key to this better future lies in science; and the faith in this future, and the exultation of contributing to its realization, have more to do with the enthusiasm of Russian youth for science (and with the recent precipitous growth of science and technology in Russia) than the carrot of material compensation or the stick of government dictation. If we want to instill a similar devotion to science in our own youth, we, too, have to offer them more than the inducement of higher salaries or appeals to the necessities of military defense. We have to make our youth aware of the state of the whole world and the wants of mankind as a whole, and of the precariousness of our exclusive national prosperity and security in the midst of a largely miserable and restless world, and arouse in our youth enthusiasm for science as a force which can help us to emerge from this dangerous predicament. This is what "education for life" and "education for citizenship" must mean in our age—education to face the life of mankind as a whole and to accept the responsibilities of a citizen of the world.

SCIENCE FOR THE NON-SCIENTIST

The social and political revolution of our time could have occurred —and probably would have occurred—even if scientific progress had not experienced the fantastic acceleration we have witnessed since the explosion of the first two atom bombs in Japan; but, as it happened, the technological revolution, superimposed over these social and political changes, has given them an entirely new significance and urgency. Americans would have looked with equanimity, if not with sympathy, on the destruction of European colonial empires and the emergence in Asia and Africa of national states in place of long-stagnant feudal societies. Contrary to the belief of Communist theoreticians, for whom the implacable hatred of capitalism for the growing socialist society is an article of faith, Americans could have—and probably would have—watched, without much apprehension, even the unprecedented struggle to impose a new form of social organization on nearly one-third of humanity by indoctrination and terror. What makes us deeply concerned is that the violent conflicts which such drastic political, racial, and social changes inevitably generate are now a threat to humanity as a whole because of the mastery which the several powers, old and new, possess (and others soon will possess) over the forces of the fissioned and the fusioned nucleus. These are forces millions of times stronger than all that were at the disposal of mankind prior to 1945. Because of this, we need, for our own survival and that of mankind as a whole, an open-minded leadership—men who will not only understand the great national and social movements and revolutionary ideas now abroad in the world but also appreciate the impact of new science and technology on human affairs. They will need to be able to project into the future the effect of these new forces on the fate of nations, since the unique thing about science and technology is that they are dynamic phenomena, developing at a rate which leaves far behind the much more slowly evolving—or stagnant—forms of political, economic, and social life. Building the future on the basis of today's science and today's technology would be even more foolish than building it on the basis of today's political and social relations.

In a free society, if it is to remain free, the requirements of leadership are also those of citizenship. A free country cannot take decisive political steps unless their meaning is understood and their necessity comprehended by the citizenry. It is the task of education to make this understanding possible for the generations to come.

Practically every major political decision of our time requires taking into account scientific facts and technological possibilities. In the words of Sir Eric Ashby: "Even members of the House of Lords find themselves called upon to make decisions about radioactive fall-out, overheating during supersonic flight, and the strontium content of bones." [2] At the present time in America, the incorporation of scientific points of view into the making of political decisions is done by a highly unsatisfactory, haphazard method. Political leaders, who have a poor—if any—understanding of science and little familiarity with technology, receive, more often than not behind closed doors, advice from a few chosen scientific and technological experts. They have no competence to properly evaluate this advice, particularly if it comes not in the familiar terms of yes-or-no, true-or-false statements but, in conformity with the nature of scientific knowledge and inference, in terms of possibilities and probabilities. (Often, to make their advice understandable, scientists feel that they have to express it—against their better judgment—in the simple black-or-white, do-or-don't form.) The expert opinion is then tossed into the big kettle in which the political brew is being cooked from the usual major ingredients: national interests and idiosyncracies, economic desires of various social groups, and party-political considerations. Scientific and technological arguments, however incontrovertible they may be, are thus diluted and often altogether lost in policy determinations. We may not go as far in overruling them as Stalin did when he endorsed Lysenko's pseudo genetics by the syllogism: communism needs rapid transformation of the species of plants and animals; genetics says this is impossible; *ergo,* genetics must be wrong. Our political leaders do not flout the advice of science so brazenly; but they do often ignore it or pick out, among dissenting scientific opinions, the ones which best fit their political plans and not the ones which carry the best scientific support.

If democracy is to be real and not a sham in the coming scientific

[2] Ashby, "Technological Humanism," *Impact,* IX (1958), 45.

age, we shall need a public opinion, and a political leadership, able to exercise sound judgment in situations involving scientific or technical facts. This will not require an extensive study of science by all citizens (or even by those intending to seek public offices) but merely a general education producing men and women capable of making a *proper choice* of advisers, according to their scientific competence and not their political conformity—men prepared to ask *proper questions* of these experts and capable of understanding and evaluating their answers, in the same way most legislative and political leaders are now capable of evaluating the advice of legal or economic experts. This is not as difficult as many seem to believe. The arguments of science are, in fact, often simpler and closer to common sense than are those of lawyers or economists. This is not to imply that politics should or could become a science in any foreseeable future. There will be plenty of scope left for imagination and political instinct, even in a world where scientific evaluation of facts and possibilities of a political situation will be accepted with all the respect it deserves; but politics as art will begin on a much higher level of solid understanding than it does now.

Of course, we do need more and better science teaching in our elementary and high schools than we have now; and this teaching should involve less attention to gadgets and more attention to fundamental facts and concepts—less play and more work. Otherwise, our future generations will be deprived of an understanding of some of the most important things around them; and the selection, from their midst, of the group (inevitably small) who will have the thrill of participating actively in advancing this understanding will become accidental and inadequate. However, what is required is not the heaping of one scientific course upon another or the stuffing of the existing ones with more and more subject matter. Scientific knowledge now snowballs so rapidly that the schools cannot hope to supply students, as they were able to do in the past, with the most important factual content of science, or of any of its major branches, as part of general education. What is needed, instead, is a careful selection of material from certain key areas of science suitable for the demonstration of the methods by which science approaches and explores nature, of the ways in which it arrives at its general concepts, of the types of questions it asks and the types of answers it receives, and of the generalizations and mathematical laws and theories it derives from these answers, so that, on leaving school, students will comprehend the potentialities as well as the limitations of the experimental method and so that the understanding of the practical utility, heuristic importance, esthetic beauty, and philosophic value of general scientific concepts and mathematical laws will be a part of their general intellectual endowment. Science education on the high-school level should leave open to every intelligent student the option of entering, if he should so desire (or if life should require it of him), a field of science, not as a bewildered and helpless stranger, but as a traveler who knows how to find his bearings in an unfamiliar country. To quote Derek J. Price:

In classical Greece, when the great growing-tip of civilization was the new concept of Democracy, there were two classes of people: those who were actively involved in this movement and those who were not. The participants were those holding public office—the others, disinclined to such office by nature or nurture, were called *idiotes.* The word proved useful and its meaning was extended from those ignorant of politics to those ignorant of other sections of civilized life—they were *idiots.* . . . Now, without implying that Democracy is one iota less important today, and without over-glorifying Science, I should like to suggest that Science has largely replaced Democracy at the growing-tip of civilization. . . . At the same time that our civilization has become so sensitive to the benefits and woes engendered by science, we have bred amongst us a new class of idiots—a class of those bereft of office, participation, or appreciation of this force which is driving humanity.[3]

The last sentence leads to the most important plea I would like to bring forth: the necessity to provide future generations not only with general understanding of science as such but, most of all, with the capacity to appreciate those aspects of science which affect the future of man—the impact of science on public affairs and on the fate of our own nation and of mankind as a whole. This means that science should be taught not as a separate body of technical facts, or an autonomous system of ideas, but in relation to other disciplines that traditionally mold the attitude of growing generations toward the society and the world they will have to live in: history, political science, sociology—even religion (although in America, the latter is kept strictly outside the sphere of public education).

This integration of scientific education into the "classic," humanistic studies has two aspects. In the first place, science should be taught with emphasis on the consequences of scientific progress for the general affairs of mankind, on the transformations which the great discoveries of science—we have only to think of evolution, or relativity, or atomic energy—have produced in man's general outlook and position in the world, on the social and ethical problems created for the scientists and for society by man's increasing mastery over nature

[3] Price, "The Scientific Humanities—An Urgent Problem" (Second American Humanities Seminar, June 29, 1957).

and its forces. The other—and to my mind more important and more neglected—aspect of the integration of science into the general education of the citizen is the reverse of the first: the incorporation of the results and attitudes of science into the teaching of the humanistic disciplines (and, if churches would be willing to listen, also into the teaching of religion!). The teaching of the humanities—in particular, of law, government, and political science (and to a lesser degree, also economics)—is traditionally static; it takes the existing state of affairs in the world for granted, the nature and capacities of man as constant. Admittedly, it is willing to consider the historical development of laws and institutions but only for the purpose of explaining the present state of affairs, to justify what there is, not to appreciate its transient nature or to try to predict how it will—or should—change. The present is shown as the culmination of the past, not as a fleeting moment in the increasingly steep climb of mankind toward its future.

Science, on the other hand, is in such rapid development that it has acquired an essentially dynamic outlook. It looks at its facts, concepts, and laws as a temporary and rapidly changeable codex, reflecting the momentary state of affairs. (It was not so in the middle of the nineteenth century, when science offered the outside picture of an almost finished, complete, and immovable edifice, based on solid foundations of Newtonian mechanics.) With the increasing impact of science on all aspects of human existence, the humanities need to share at least some of this dynamism, this anticipation of change and the readiness to consider it with an open mind.

Science places man in a vast perspective of space and time, and this relation needs to be reflected in man's attitude toward himself, in his concept of his status and his importance in the universe. This does not destroy the legitimacy of a man-centered philosophy or a religion postulating man's importance in God's scheme of things; but it places both in a new light. It teaches man a new kind of humility.

The scientific approach to the world around us is exploratory, tentative, relativistic, quantitative; it constructs temporary working hypotheses and discards them; it expresses its conclusions in terms of relative probabilities, not absolute certainties. The humanities, on the other hand, are accustomed—in particular in the study of law, government, and social and political sciences—to qualitative arguments and to absolute concepts. In the words of George Boas: "Humanities have remained, on the whole, slave to what their practitioners believe are the universal and eternal values." [4] In my presence,

[4] Boas, "The Problem of the Humanities," *Journal of General Education,* X (1957), 205.

a world-renowned lawyer argued that blood tests are inadmissable as evidence in paternity cases because they carry a definite, even if small, probability of error (due to mutations); court verdicts, he said, cannot be based on "probability" but only on "certainty" of the judge's or the jury's conviction, which, in his opinion, can be derived, in paternity cases, from the testimony of witnesses or the impression left by the plaintiff!

Mankind will need more of the open-minded, tentative, statistical, relativistic approach to many supposedly absolute truths or values in the life of individuals or society if it is to survive and prosper in the scientific age. Science has shown the way with the relativization of such former absolutes as space and time, matter and energy. Bohr's complementary principle has taught science how to use two, on the first glance, mutually exclusive concepts—the wave and the particle—as two equally valid, parallel, and not contradictory approaches to the same physical reality. He has thus opened the door to an intellectually satisfactory justification for the coexistence of concepts, such as causality and freedom, which have been contending for absolute supremacy in our philosophy of history and human behavior. A quantitative, relativistic, complementary approach, instead of the postulation of qualitative, absolute, unique truth, is what modern science calls for in our relation to state and society. To quote Joseph Gallant: "What is needed is a reorientation in our intellectual outlook and in our literary and artistic output, accompanied by a reorientation in the teaching of arts and of the humanities, to gear them to a scientific outlook." [5]

Humanities for the Scientist

The need for better understanding of science by the citizen who does not make science his vocation (or avocation) is paralleled by the requirement that scientists should have more education in the broad field of the humanities. The scientist is often represented as a narrow specialist with little understanding and interest in literature and art, in history and social problems, and utterly naïve and gullible in his political convictions and actions. This is the standard picture of the scientist, not only in comic strips or television plays, but also in literature, even when the author is as familiar with the scientific milieu as Aldous Huxley should be by his family background. It is easy to point out that this picture is inapplicable to many scientists, in fact, to a large majority of them; that many, particularly the really outstanding ones, are very widely read; that the most abstract

[5] Gallant, "Literature, Science, and the Manpower Crisis," *Science,* CXXV (1957), 787.

of them, the theoretical physicists, are, as a rule, highly musical (Planck invented a new musical instrument; Born, Heisenberg, Teller are accomplished pianists; and Einstein's violin is a familiar attribute). Even in the field of politics, scientists are not as naïve as they are often supposed to be. When many of them propose radical solutions for the problems of our time, it is not because they are unfamiliar with the real state of affairs in the world and do not appreciate the difficulties of overcoming the traditional, ingrained prejudices and irrational passions which largely determine the behavior of men and nations. They are well aware of the difficulty of applying rational methods to the solution of human problems; but they are even more acutely aware of the consequences of not trying to overcome them—a necessity which many non-scientists simply refuse to face.

On the other hand, it is true that, with growing specialization and a rapid increase in the number of scientists, science ceases to be the occupation of a few exceptional minds, many of whom naturally tend to encompass, together with their concern—or even passion—for scientific knowledge, the enjoyment of art and literature, and a compassionate understanding of the social and political problems of humanity. Instead, we are now developing thousands of scientists from average, or slightly better than average, students. These young men and women are under pressure to concentrate their attention on technical studies to keep up with the hard requirements of professional success in their chosen field. Instead of the purely intellectual satisfaction, which has been the main lure of science in the past, we are rewarding these large cohorts of young scientists with attractive salaries and (at least in future prospect) with social prestige and influence. Our security and loyalty policies encourage them to stifle their political interests. Under these conditions, there is a real danger of creating a new breed of scientist—a technician, with a mind narrowly developed in a single area, who is satisfied with professional success and who does not have broader intellectual interests. The need is real to inject into the educational curriculum of this new generation of scientists topics of general humanistic interest, to prevent their one-sided development and to make them into harmoniously developed individuals, capable of appreciating man's achievements in fields other than science and engineering and aware of the historical developments and social problems of the society in which they live.

Attempts to enrich the education of scientists and engineers have been under way, in recent years, in our major engineering schools. With the return of high schools to more intensive education in basic subjects, such humanistic studies on the college level will become more productive. We can hope that future American scientists and engineers not only will have taken college courses in literature, art, philosophy, and sociology, but will have learned in high school how to express themselves in clear and grammatically correct English, with an adequate vocabulary and the ability to use various punctuation marks in addition to full stops, and that they will have an idea of the location of different countries on the map and the general outline of their history.

This broadening of the education of scientists is, however, not the aspect of the problem which I would like to emphasize today. Rather, I am concerned with the teaching of science itself as a humanistic subject, as I was concerned above with the incorporation of the facts, concepts, and methods of science in the teaching of the humanities. I have argued that in the scientific age no man will be able to call himself properly educated in history or the social sciences, or fit to act in human affairs and the development of national policies, if he has no understanding of the general viewpoints and methods of scientific thought; if he has not incorporated into his approach to human existence, as well as to national and international relations (and even to his relation to the higher moral forces of the universe), the characteristics of open-mindedness and tolerance, the principles of relativity and complementarity which have been evolved by modern science and have made it so successful in its dealings with the world around us. In the same way, science cannot remain an activity unrelated to the humanistic principles and values which mankind has evolved in its history. Scientists should be able to see science as part of a broad human striving for the achievement of a better and more perfect state of man. It is not enough to teach the scientist some philosophy, some history, and some sociology or to awaken his interest in art. He has to learn to integrate science with the humanities and not to place the two mechanically one beside the other.

I can do no better than to quote here from Sir Eric Ashby, the scientist-chancellor of the University of Belfast:

> Ninety years ago, when Thomas Henry Huxley was giving evidence before a Select Committee on Scientific Instruction, he poured scorn on the universities because they made literature and grammar the basis of education and (as he put it) "they actually plume themselves upon their liberality when they stick a few bits of science on the outside of the fabric." It will not do for us, ninety years later, to reverse the process and to plume ourselves upon our liberality when we stick a few bits of the humanities on the outside of the fabric of higher technology. That would not achieve our purpose; it would only generate in students a contempt

for the humanities, analogous to the contempt which public schoolboys in the 1860's had for science. It is no good trifling with this problem. We must cut the path to a liberal education through a man's specialism, not by by-passing it for an hour or two a week.

It should be our business in higher technical education to persuade students that they cannot practice technology without continually reflecting upon its social implications, and that some understanding of the humanities is essential to this reflection. For a metallurgist to take a course in general history in order to acquire a little culture is an innocent but not necessarily a very profitable amusement. But if the metallurgist studies the social and economic history of the Industrial Revolution; if he learns how the making of iron followed the forests across Europe until coal replaced charcoal; if he realizes how mining and the building of railways and canals stimulated the study of geology, and how this in turn brought men face to face with the evidence for organic evolution; if he considers the effect which the introduction of electric power has had on the distribution and layout of factories, and the effect which aluminum alloys have already had (through world transport) on the conduct of diplomacy: then history becomes integrated with metallurgy; his humanism becomes a relevant part of his day-to-day thinking. . . .[6]

Later, Ashby quotes Samuel Alexander and comments:

"Liberality in education is a spirit of pursuit, not a choice of subject." If technologists are to enter fully into their responsibilities in the world of tomorrow, they have to adapt themselves not only to novel technologies: they must also be able to adapt themselves to the social consequences of the novel technologies. Their prospects of adaptation certainly depend in part on the spirit with which they pursue technology as undergraduates. To pursue technology in a spirit of liberality, they need not only a thorough training in the theory and practice of technology (they get that already); they need also an understanding of men and communities such as is acquired through literature and history and the social sciences. . . .

All industrial peoples have to solve a great problem in human ecology, namely how to adapt themselves, through the second channel of inheritance,[7] to the social climate which is being created by modern technology. Technologists have a special responsibility to solve this problem for themselves and to reflect on its implications for their fellow citizens. I am persuaded that the solution for technologists lies through the inheritance of education they receive, and I am persuaded that the appropriate education for higher technology should be one which does for contemporary civilization what our traditional classical education has so successfully done for the civilizations of Greece and Rome.

[6] Ashby, *op. cit.*

[7] In Ashby's terminology, the "second channel of inheritance" is the written and the spoken word, while its "first channel" is provided by the genes.

George Boas recently commented about scientists and technologists called in as experts to assist in policy decisions:

> Since I am neither a scientist nor an engineer, I do not know what takes place in their offices when they are in conference, and hence have no idea whether they ever think of the humanistic consequences of their plans or not. But I judge from talking to those whom I know that they conceive of their problems as cut off from all such subjects. This was exactly the reflection of the Nazi war criminals when shown in films the results of their orders. It is this technique of chopping up problems into specialties which has created the problems of the humanities. If one thinks of these studies as meditations on the philosophic importance of facts, then the chopping is clearly absurd. By facts I mean, of course, descriptions of the universe as given us by the sciences, whether these sciences be physics and chemistry or biology and psychology. And by meditations I mean wondering about such problems as how men can achieve their ideals if the universe is as described by the sciences. . . .[8]

In 1944, with the approach of the development of the atomic bomb to its successful culmination at Alamogordo, a stirring of responsibility took place in the minds of the scientists associated with this development. Some of them petitioned the President not to use the bombs in Japan for moral reasons; others wrote memorandums to the government in which the same conclusion was drawn by analyzing the probable influence of atomic energy on the political and strategic developments of the postwar era. These scientists anticipated the rapid development of the atomic arms race, the creation of "super bombs" a thousand times more powerful than the bombs used in Japan, and the ultimate domination of the whole international situation by the threat of nuclear war. They argued that the discovery of atomic energy should become the signal for the beginning of a "new deal" in world history—for a resolute break with the age-old tradition of wars as legitimate means of solving international conflicts. A new age of world community and world law must be inaugurated, not out of "unrealistic" idealism, but out of the intelligent interest of all nations in self-preservation. Moral revulsion against the indiscriminate mass murder inevitable in the use of atomic weapons merged in this way with the conclusions derived from objective extrapolation of traditional, prescientific history into the atomic age.

For a while, after the secret of the existence of the atom bomb had been revealed by the holocausts of Hiroshima and Nagasaki, it seemed that the realization of the from-now-on inescapable involvement of scientific activity in the affairs of mankind, and the consequent responsibility of scientists for the future of man, was spreading among

[8] Boas, *op. cit.*

the scientific community. Large numbers of scientists took part in the campaign for the civilian control of atomic energy in the United States and for the international control of atomic weapons through the United Nations. However, when the first aim was achieved and the second revealed itself as elusive, this interest subsided, and it soon became clear that the whole movement had been carried forward by only a small number of dedicated individuals.

As time passed and the impact of the first atomic explosions wore off, as the requirements of professional life became more pressing and the lure of rewards of success in science stronger, the tide ebbed. Occasionally, a flareup occurred, such as the recent declaration of seventeen German physicists against the atomic armament of Germany or Dr. Linus Pauling's petition to the United Nations against the continuation of nuclear bomb tests, signed by over two thousand American scientists. However, both these actions were in the nature of protests rather than constructive contributions to the resolution of the tragic involvement of science in the fate of a divided humanity. One could sign these documents and then go back to one's scientific work; and this is what the majority of signers did. Twelve years after the start of the movement of "atomic scientists," the majority of Western scientists have only a vague feeling of their new collective responsibility to mankind, and only a few among them are willing to give a substantial fraction of their thought, of their time, and of their energy to attempts to discharge it.

This is not enough. Scientific education in the future should have as its aim not only the production of well-trained, competent scientists but also the instilling in their minds of a concern with the implications of science for human affairs and the fostering a feeling of their coresponsibility, as scientists, for the fate of mankind in the scientific age.

It has sometimes been suggested that the teaching of science in historical perspective would provide an essential contribution to an understanding of the involvement of science in the general progress of mankind. (Eric Ashby suggested this in the above-quoted article.) This is true, but only to some extent—to the same extent to which the teaching of history of science as part of the general history course can assist non-scientists in the realization of the essential role science has played in the progress of human society. In both cases, much more than the knowledge of historical developments is needed. In high-school teaching, the ethical and moral aspects of science, the relation of scientific truth to the general system of human values, and the responsibility of science and scientists for the future of society must be impressed on students. On the college level, scientific education should build up from these foundations and produce scientists who are citizens in the full sense of the word, able and willing to accept not only the general duties free society imposes on all citizens but also the special responsibilities placed on them through the decisive participation of science in the present and future history of mankind —in fact, in the determination whether this history will continue at all.

I did not touch in this paper on the contributions of the sciences (or the humanities) to the intellectual enjoyment and spiritual achievement of man as an individual but only on their contributions to the improvement of human society. This is not because I deny that the highest reward of all education is the capacity it can bestow—or at least enhance—to commune with great spirits, to share great ideas, to partake of the joys of spiritual and artistic accomplishment, and to experience the thrills of penetrating into the inner workings of nature and of unravelling the mysteries of life and mind (and—the highest delight of all—to contribute creatively to these achievements, which determine man's spiritual stature in the world). I have abstained from talking about these matters because I am overwhelmingly conscious of the threat which, in our time, the deficiency of human society poses to the very survival of man. It raises the danger of losing all of mankind's past achievements and destroying man's capacity to add to them or replace them. Man as biological species is in danger of going down the drain of natural history as an unsuccessful experiment, despite all the saints and philosophers, poets, artists, humanitarians, and scientists he has produced. In the face of this danger, I can see no answer in the development of more perfect individuals. The God of the Bible was willing to preserve Sodom for the sake of ten righteous men, and the Roman Catholic church sees the purpose of mankind in the production of saints. Being a scientist, I cannot see the fate of mankind in our time as depending on spiritual heights reached by a few individuals. The crisis mankind is facing now will not be resolved by the increased capacity of individuals for intellectual achievement or artistic creation but only by the advancement of human society from its traditional semibarbaric state to one adequate for survival in the scientific age. It is under this most urgent aspect that I have considered the roles of science and humanities in education—not *sub specie aeternitatis.*

27

Integral Science and Atomized Art

The evolution of the human mind is a single process, revealed with different intensity, different clarity, and different timing in its various manifestations—are, science, philosophy, social and political thought. It is like a fugue, or an oratorio, in which different instruments, or different voices, enter in turn. The voice of the artist is often the first to respond. The artist is the most sensitive individual in society. His feeling for change, his apprehension of new things to come, is likely to be more acute than that of the slower-moving, rational, scientific thinker. It is in the artistic production of a period rather than in its thinking that one should search for shadows cast in advance by coming events, for prophetic anticipation. I do not mean the forecast of future events but rather the revelation, in the framework of artistic productions, of the mental attitudes which only later will become apparent in other fields of human endeavor. Thus, the impending breakdown of the existing order of things, of the generally accepted system of values, should be—and often is—first recognizable in a revolt against the values and canons that have dominated artistic creation; a revolution in art precedes the revolution in society. In the same way, the stabilization of a new political regime, or social order, is often presaged by the acceptance of new canons in art, be it the glorious neoclassicism of the French revolution or the drab "socialist realism" of the Russian one.

The reaction of art to an impending change is revulsion rather than eager anticipation. When science first began to change the lives of men, many artists revolted violently before the vision of a future irrevocably committed to science. This is a futile revolt: science is the only field of human activity in which progress is incontrovertible and inevitable. Artists, politicians, philosophers, can look back to this or that period in past history as a peak of human achievement, be it classical Greece or Renaissance Italy, the civic order of the Roman empire, the solidity of Victorian England, or the golden eras of Indian or Chinese history and art. No such nostalgia is possible in science. Its highest point is the present, and every successive year is bound to be an advance beyond the preceding one. A civilization

From *Bulletin of the Atomic Scientists,* February, 1959.

that refuses to accept this constant progress and to adjust itself to it does so at its own risk; and that risk is, at the present, appalling.

In 1835, the Russian poet Baratynsky had the dismal vision of the "last poet" arising amidst a generation "marching on an iron path" and vainly trying to awaken this generation, "coldly devoted to industry and profit," by "simple songs, praising love and beauty and revealing the emptiness and vanity of science that flouts them." In another poem, looking longingly backward, Baratynsky praised the times "when man had not yet tested nature with furnace and scale and compass." And in France, about the same time, Lamartine deplored the "heavy compass of Euclid that stifles our enchanted arts." He accused the generation of his parents—the Age of Enlightenment—of having "replaced the transports of the soul and of the genius by cold passion for calculations," of having "congealed nature with the icy finger of measurement."

Only rarely has a poet or a painter been excited by the feats of science, and when he was aroused, it was more likely to be by the glory of martyrdom for science than by the glory of science itself. When Alfred de Vigny wrote his "Bottle in the Sea," when he found resounding words to describe the tolling of church bells and the roaring of guns commemorating the "heroes of science rather than those of battle," he was paying tribute to explorers, "soldiers of science" facing the blind fury of nature, not to the soaring of the creative scientific mind; to the deeds of a Magellan or a Captain Cook, not of a Euclid or Newton, Copernicus or Pasteur.

Two Aspects of Science

The revulsion of so many artists and poets against science is easy to understand. They considered science a cold, analytical tool, dismembering the beautiful wholeness of nature and reducing it to a heap of meaningless parts. Science, they believed, destroyed things to better know their elements and, in doing so, took the soul and the spirit out of them. It was this "atomistic" spirit of science that impressed (and still impresses) most non-scientists and offended many a sensitive artist. He did not—does not—realize that science is also the mainspring of the most daring and revolutionary generalizations, universal concepts and comprehensive hypotheses, giving new meaning to the universe, permitting man to contemplate the world around him as a whole with a new and deeper admiration and awe.

It is, on the other hand, just this synthetic aspect of science that has inspired those rare artistic geniuses who have been deeply attracted to science—a Leonardo da Vinci, with his admiration of the universal laws of perspective, with his feeling that the same structural principles underlie the efficiency and the beauty of the living body and of the machine, or a Goethe, with his intoxication by the metamorphosis of plants and animals, the vision of a universal law underlying the beautiful variety of living forms. Goethe, like Lamartine, worshipped nature; but whereas for Lamartine measuring and analyzing nature meant destruction of life and beauty, for Goethe it meant revelation of a new and deeper beauty. And that is what it meant—and means—for many, perhaps for most, great scientists.

This is particularly true of our time, when the dominant theoretical structure of the nineteenth century, Newtonian mechanics (which was largely responsible for the picture of science as a cold, analytical tool, mechanizing and despiritualizing the world), has been complemented by other breath-taking generalizations, suggestive of deeper interrelations and unities in the sensual world: Maxwell's electrodynamics and Einstein's relativity, quantum statistics and electrodynamics, and—just outside our present reach—a unified field theory incorporating all these concepts. Penetrating deeper and deeper into the world of almost infinitely small subatomic particles, science also approaches the understanding of the almost infinitely large world of cosmic, astrophysical phenomena and relations—the mystery of the small is also that of the large; analysis is a step to synthesis. Of that, the non-scientific world has hardly any awareness, and neither have the creative artists. Here lies, perhaps, a fatal shortcoming—one fatal shortcoming—of our science education: it is much easier to describe how phenomena are taken apart than to show how the separate facts and observations fit into great general laws and concepts of which they become but special cases. Religion, philosophy, art, remain in the eyes of most people the only legitimate sources of general ideas concerning the world and man—ideas still barely tinged by all that science has revealed in the last hundred years.

Atomization in Art

When art did succumb to the penetration of science, it was to that very aspect of science it most violently denounced—the dismemberment of form, the atomization of matter, the taking apart of phenomena. Artists, musicians, and poets joined in the destruction of visual, musical, and verbal wholes which men for ages had learned to cherish as expressions of an inner unity and integrity, endowing them with a spirit or soul. The wholeness of the human form, of an animal, a plant, a building, or a landscape, was destroyed and replaced by a conglomeration, on the canvas or in space, of simplified, idealized, elementary building stones—simple parts into which the living entity was decomposed.

At first it was but a playful, superficial decomposition. The pointillists found that people, or a summer landscape, could be represented by an iridescent interplay of colored dots; but the viewer, after overcoming the initial shock, could still see the momentarily hidden entities. It was merely a transparent, dazzling cloak thrown over the familiar solid, three-dimensional forms. The change became more serious when decomposition into elementary units went into depth. When Cezanne first took mountains apart and replaced them by stone heaps and Picasso and the cubists decomposed human bodies into blocks, a decisive step was made from "synthetic" to "analytic" art. The next step was to scatter the heaps apart—and then, in an ultimate simplification, to flatten them into squares and polygons, ending with the multicolored geometrical patterns of a Mondrian. This seems to me the ultimate in the influence of science on modern painting (and analogous things have happened in music and poetry!), with science conceived as reduction of the whole to parts endowed with autonomous life; the ultimate refuge of art becomes the arbitrary arrangement of these parts into dazzling patterns.

In modern art one does not find science conceived as a path to spiritual universality, as a means of establishing relations between previously apparently independent entities; one does not find the thrill of sensing the beautiful organization of the world as a whole along simple, logical lines. I do not know how this synthetic, unifying spirit could express itself in art. Certainly not through the adoration of simple circular shapes or spherical or ovoid forms, as in the sculpture of a Brancusi. But some works of a Moore or an Epstein do evoke in me the feeling of this universal and simple unity. Only creative artists could say (or, rather, show in their art) how the beautiful simplicity of Maxwell's field equation, Einstein's relativity concepts, or Bohr's quantum conditions can find a counterpart in the analogous task of art to interpret the world to man in an entirely different medium. When these works of art are finally created, it will become obvious to all how it can be done.

Non-Objective Art

When asked to explain why they paint in the way they do, instead of representing known forms and objects, abstract painters often say that representational painting has been made superfluous and artistically impossible by the perfection of photography; that the exact representation of things as we see them must now be left to mechanical and optical arts; that the latter can not only reproduce the exact forms and colors but even endow them—by proper illumination and other tricks of the trade—with various moods and harmonies. This, however, is not a valid answer. Mechanical or optical tricks could probably also imitate cubist or abstract painting, distorting and decomposing original objects until they were hardly recognizable. On the other hand, no photography can ever approach the golden shimmer of an interior by Vermeer, or the mystic depth of a night scene by Rembrandt, or the rigid splendor of a cavalcade by Mantegna, or the velvet softness of a painting by Giorgione, not to mention the stylized beauty of Japanese landscapes or the grace of Indian dancing gods. It is not true—it cannot be true!—that *all* possibilities of modifying, stylizing, or idealizing reality through the vision of an artist have been exhausted as art has evolved from cave painting in the Pyrenees through the marble goddesses of Hellas to the blue and golden paradises of Fra Angelico and the flowery iridescence of Renoir and Degas. Every time our knowledge of art has spread in breadth, taking in new areas of the world, it has discovered new ways of transforming reality into new shapes and colors, evolved by a different race but representing the same real things of the world. There must be an endless number of ways to see the physical world and endow it with beauty, meaning, spirit, and soul, without destroying it into atoms; and future generations of artists will discover them. And the same applies to the representation, in poetical words, of the events in human life, of man's sensual impressions, and of his thoughts and passions.

Non-objective art is often rationalized by analogy with music, where the occasional requirement of descriptiveness is not taken seriously, not being satisfied by the greatest masterpieces with the most enduring appeal. In long centuries of musical development, man has learned to respond to the flow and interplay of sounds only vaguely reminding him of the realities his eyes or ears perceive in nature. These abstract musical designs, consonances and dissonances, acquire the significance of something of utmost beauty and importance. Cannot the same become true of the flow of graphic design, of color dissonances and consonances in an abstract painting, if the sensitivity of man for this kind of beauty has time to develop, so as to match his musical sensitivity? There is one great obstacle. In painting, or sculpture, one can see a counterpart to harmony, but there can be only a vague suggestion of melody. The power of music lies in its mastery over the fourth dimension, the dimension of time. It is through the melodic curve, through crescendos and diminuendos, counterpoint, and other developments of musical phrases and movements that a work of music acquires its wholeness, its unity, all that makes listening to it a great experience. This dimension is missing in graphic arts or sculpture, unless the future develops new kinds of art in which shapes and colors will be displayed not statically, but kinematically, in time (and, more than that, dynamically: that is, conveying the feeling of logical necessity, of inner forces which make the changes of time and pattern convincing and not mere arbitrary variations).

Without this fourth dimension, abstract art is not likely to acquire a significance for the beholder equal to that of a musical piece, a poem, or a picture made significant by its content. At best, I believe, it can become as aesthetically satisfying (and even beloved) as a decorative piece—a Persian carpet or a porcelain vase from China. This is not to disparage the beauty of these products of applied art, in which whole nations and whole eras of human history have been preserved for the admiration of other nations and other generations; but there is, nevertheless, a wide difference in the relative importance, in the inventory of world art, of even the most gorgeous Oriental rug or the most exquisite porcelain vase and of a great painting, poem, or symphony.

Integration of Mankind

In science, we are familiar with the progress from the differential to the integral; from Newton's differential equations of motion to the integral concepts of dynamics, to Hamilton and the quantum-mechanical integrals; from the particle to the wave and the field. In its application to social and political problems of the world in our time, science becomes a force for the unification and integration of mankind. It sees mankind as a single organism rather than as a conglomeration of individual cells and molecules following their own paths—self-centered nations and classes—every one in pursuit of its own interests. The fate of mankind depends on whether it overcomes its atomization, inherited from the prescientific age, or persists in it. Perhaps one should look at art, if one wants to anticipate whether the liberation of mankind from its own "atomic age" is at hand. The consciences of men everywhere show signs of a change from exclusive concern with the interests of relatively small economic, political, and national associations toward wider concern with mankind as a whole. Despite the persistence of old, and the blossoming out of many new, particularisms, the feeling that all mankind is in one boat, to sink or to swim together, is growing before the specter of a common nuclear catastrophe. Art may be the mirror and the barometer revealing and anticipating the struggle between the forces of atomism and fractionation of the human conscience, human passions, and human society and the creative forces of harmony and integration. Not through its choice of subjects or its exaltation of this or that idea, but entirely in its own medium and through its own means of revealing, liberating, and giving form and expression to the spiritual stirring of man, could art not only anticipate but also assist mankind in finding an answer to the challenge posed by science!

The Atomic Age Doctrine

National Interests—The Traditional Basis of Foreign Policy

The factual framework within which the foreign policy of all countries, including the United States, must be laid out has been changed beyond recognition by the scientific revolution of our time. Traditionally, each sovereign nation has had the right—and many thought it a duty—to conduct its foreign policy primarily with the aim of furthering its own national interests, and this is what all nations did most of the time, with occasional proclamations (and sometimes halfhearted pursuit) of more general ethical, religious, or economic principles. Smaller or more exposed nations often could not effectively pursue their interests except by banding together in alliances; but nations as powerful and naturally protected as the United States could carry on a self-centered policy on their own. *National security*—meaning by this the protection of home territory and population from invasion or destruction in war—appeared easily attainable, if certain measures of military preparedness were taken; and *national prosperity* appeared to be a subject of foreign policy only to the extent of protecting home production from foreign competition through appropriate tariff policies.

It is easy to adduce proof that this concept was not adequate even before the technological revolution of our time. The step-by-step involvement of the United States in two world wars and the spread of the economic depression, originating in America in 1923, to the rest of the world suggest the degree of strategic and economic interdependence of nations and continents that already existed prior to this revolution. Nevertheless, the traditional framework of American foreign policy could and did remain centered around our national interests in the direct, narrow sense of this term until the Second World War; in fact, this appeared to be the only realistically conceivable framework. Many still conceive it as such.

However, the technological revolution has changed this situation radically. This revolution has shrunk distances until every part of the globe can be reached from every other part—by man in one or two days, by missiles in one or two hours, and by messages in a fraction of a second. This in itself calls for a new look on the possibilities and limitations of the traditional self-centered national policy. The needed revision would have been less drastic if the technological

A paper written in 1960.

revolution had not also produced an enormous change in the amounts of energy available to man. Not only can we transport men and objects thousands of times faster and send messages millions of times faster than was possible a century ago, but since the discovery of nuclear energy, we can also carry or send over wide distances energy packages millions of times greater than our fathers had at their disposal twenty years ago. The simple and stark result of this development is that war has ceased to be a rationally acceptable arbiter of international conflicts, an *ultima ratio regis,* and has become a nightmare of national murder-*cum*-suicide.

Yet, as long as a number of sovereign states pursue foreign policies determined primarily by their national interests, conflicts between these interests inevitably arise. Some such conflicts were, in the past, adjusted by compromise, whereas others were resolved by war. What is important is that, even when war was avoided, its possibility was a decisive factor in reaching a compromise settlement. The military power of a nation, and its readiness to use it rather than acquiesce to policies of other nations deemed detrimental to its national interests, stood as an invisible second behind the diplomatic fencer in the arena of world politics. This traditional mechanism of resolving power conflicts by actual or threatened use of force is now breaking down. Diplomats still use threats of war in wrangling over conflicts of interest (such as the fate of Berlin), but these threats have lost their rationality and credibility. Therefore, controversies remain unresolved; conflicts are frozen by fear of a nuclear war without being reduced in acerbity by this postponement. And despite its irrationality, war remains an ever present possibility, not as the result of a deliberate political decision, but as an act of national desperation or as a catastrophe, unwanted by all, which could follow from the inertia of power politics or a tragic technological mishap.

New Aims of National Policies

While the old mechanism of settling international conflicts by force or threat of force is being rendered inoperative, the obvious alternative—conflict settlement by a legal body acting on the basis of acknowledged law and able to enforce its decisions—remains unavailable because of the absence of all three of its essential ingredients: an impartial international body, acknowledged law, and the power of enforcement. This fact causes many to seek a remedy in the creation of new (or reinforcement of existing) international institutions and their endowment with power attributes needed to make their decisions enforceable. Strengthening the United Nations, abolishing the veto power in the Security Council, extending the jurisdiction of the World Court, creating an international police force under the UN, are among the suggested remedies, and at the end of the path, there looms some kind of world government.

This approach seems logical; and yet, I believe it to be unprofitable at the present time without a previous radical change in the attitudes of nations and the acceptance of new principles as to the aims of national policies. No body formed by representatives of the several national states, which are dedicated to the pursuit of their own interests, can qualify as an impartial juridical body to adjudge conflicts between such nations; no law promulgated by a majority vote, or even by unanimous consent, of such representatives can be acknowledged as supranational law, and we cannot expect military power to be placed at the disposal of such a body to enforce its decisions. This situation can only be changed if the main purpose of the nation states ceases to be the pursuit of their national interests and becomes the pursuit of common interests of mankind. Only peoples and governments dedicated primarily to the universal good of mankind (and only secondarily, to the fostering of their own group interests) can aspire to the creation of a world body whose objectivity in international conflicts and supremacy over national governments has a chance of being universally recognized and which could therefore be entrusted with superior military power.

If this is so, then the most promising way to the solution of the crisis of our time leads through a change in the purpose of national governments, and this in turn requires, in particular in democratic countries, a change in the conscience of their peoples. This seems to be a roundabout and slow way, but it may be the only realistically feasible one. This approach has the further advantage that every nation can choose it and advance along it without waiting for other nations to follow suit. The American people can decide that they want their government to make the security and prosperity of mankind as a whole the ultimate purpose of its activities and that this purpose should also determine the day-by-day policies of the government. Without renouncing its function as protector of the national interests of the United States, the American government can subordinate the pursuit of these interests to a wider paramount purpose, in the recognition that, in the scientific age, our national interests are in the long run best served by the progress of mankind as a whole.

The political, ethical, and even religious attitudes of mankind have been traditionally attuned—and remain largely attuned—to life in a divided world society, shaped by the pursuit by tribes or nations of their separate and often contradictory group interests. Loyalty and devotion to one's nation, its support in all conflicts with other nations to the point of sacrificing one's life for it, have been, and still remain, the cornerstones of the social value structure of all nations, capitalist and Communist alike. The power policies of a nation are not to be questioned by its citizens—"My country, right or wrong," "Political battles end at the waterfront," and so on. Some of this attitude, of course, also prevails in smaller groups, regions, and states in the United States—and in local, religious, or economic groups everywhere—but on the whole, mankind makes loyalty and devotion to one's state the highest political virtue of every individual, man or woman.

Only exceptional religious thinkers or humanist philosophers have recognized in the past the obligation of each man to mankind as a whole and have called for subordination of national interests to the wider interests of humanity. Within the last two decades, in direct consequence of the scientific revolution, this formerly ridiculed "idealistic" attitude has become highly realistic; in fact, it may now be defended as the *only* realistic basis for a promising foreign policy. In the scientific age, continuing pursuit of selfish national interests by a number of states will inevitably lead again and again to conflicts of interest and thus to the possibility of war as a logical "continuation of foreign policy by other means." Yet, every political leader is from now on bound to shrink from war as from ultimate madness. The negative attitude of a large part of British public opinion at the time of the Suez crisis is indicative of the spreading awareness that, in our time, considerations of national interest cannot remain paramount in the determination of national policies. America may as well adjust its thinking to this reality of international life in our age, without awaiting such shocking lessons.

ECONOMIC PROGRESS OF MANKIND AS A WHOLE

What already has become true of national security is also becoming true, although in a less spectacular way, of national prosperity. Prosperity for any single nation is becoming impossible, in the long run, except within the framework of healthy economic development on a much wider, ultimately a world-wide, scale. The gradually spreading realization of this fact is due, in part, simply to our increasing (and increasingly instantaneous) knowledge of what is going on in the world, revealing the previously obscure interaction between the economic developments in America, Asia, Africa, and

Europe. The same change in the speed and ease of communication is also responsible for the realization by the peoples in the underdeveloped countries of the world of the relative inadequacy of their economic status and of the availability of a more satisfactory life ("the revolution of rising expectations").

The United States now uses a part of the natural resources of the world—metal ores, oil, fibres, and so on—out of all proportion to the size of its population; this disproportion cannot be prolonged indefinitely and serve as a permanent basis of our future prosperity. Our aim cannot be to secure for the United States permanent control of most of the natural resources of the world; instead, we must search, in a co-operative effort with other nations, for ways leading to the most efficient utilization of the natural resources of the planet as a whole for the common benefit of mankind as a whole. The same applies to food; our food surplus problem as well as the food shortages in other large areas of the world appear with increasing clarity as parts of a common problem of adequate provision of food for mankind as a whole by the efforts of world agriculture as a whole.

Since the enunciation of the Marshall Plan and the Point Four program, American foreign policy has become *de facto* involved in a program of fostering economic progress everywhere, but we are still reluctant to recognize the fundamental significance of this new beginning, which so many in this country still consider as a temporary emergency venture from which we can and should extricate ourselves as soon as feasible. At the same time, however, rapidly increasing world travel and communication brings home to an increasing number of Americans the realization of the community of their interests with those of other peoples, not only in the preservation of peace, but also in securing world-wide economic progress.

The scientific and technological revolution, which makes national security impossible except in the framework of universal security and national prosperity impossible, at least in the long run, except in the framework of world-wide economic progress, also provides, for the first time in history, a rough quantitative understanding of the present status and the extent of the technological needs of mankind and of our capacities to foster their fulfilment. We are aware both of the enormous difficulties and costs involved in a serious effort in this direction and of the ultimate feasibility of building a stable and slowly advancing world community by constructive application of science and technology. Mankind's past history has consisted largely of fighting for the control of a greater share of the natural resources of the earth—manpower, fertile lands, waterways, minerals, fossil fuels. This fight has now to cease, under penalty of an all-destructive nuclear war, and to be replaced by a conscious effort to create new wealth by full utilization of the creative possibilities of science. Again, we find a symbolic example in what happened in England at the time of the Suez crisis: frustrated in its attempt to secure by force its access to Mideastern oil, Britain decided to double, and later to treble, its atomic power program!

The foreign policy of the United States can acquire true dynamism only if it fully recognizes these consequences of the scientific revolution, identifies our own higher national interests with the interests of mankind as a whole, and resolutely assumes leadership in a world-wide constructive effort. At present, our policies are bedeviled internally, and made suspicious externally, by their lack of clear dedication to the preservation of world peace. We still behave as if we accept war with the Soviet Union as a rational alternative to maintainance of peace, whereas in fact, our military policy—for example, the increasing reliance on long-range weapons of mass destruction and neglect of tactical operational units—makes sense only as an attempt to *preserve peace* and will have failed if war should actually break out and we were forced to use them in "retaliation" for a destructive attack on our own country. It is even less clear to the world (and to ourselves) whether our assistance to nations in Europe, Asia, Africa, and South America is a measure of self-protection, an emergency policy intended to hold together an anti-Communist military alliance and to prevent the spread of communism in consequence of economic distress, or the expression of our dedication to the interests of mankind, of the realization of our irrevocable stake in the economic advancement of mankind as a whole. This is what makes our technical assistance suspect, and often little appreciated by its recipients.

An Atomic Age Doctrine

I suggest that the proclamation of an "atomic age doctrine"—the doctrine of mutual interdependence of all nations, of their common interest in the preservation of peace and world-wide economic progress—would be an effective first step toward regaining American leadership in the world. America has held this leadership twice, after the first as well as after the second world war, and has twice thrown it away by returning once to isolationism and later to traditional power politics. Such a declaration, made not incidentally in a propaganda speech but in a state paper issued after proper deliberation, would establish a stable and convincing basis on which wise and forward-looking policy decisions could henceforth be made. Many of these policies are already being carried out, but in the absence of fundamental aim-setting, their success is frustrated by furtiveness, by their alleged temporary, emergency nature, by lack of conviction on our side that these constructive policies, and not the continuation of the traditional game of power politics (from which of course we cannot escape overnight), represent the fundamental orientation of our foreign policy.

29

Science, Scientists, and International Policy

Science—An Alien Force?

At different times in history, various social, economic, and professional groups have exerted a particularly strong influence on national attitudes and the decisions of governments. From the prophets and lawmakers of ancient Israel through the philosophers of classical Greece, the civil servants of China, and the jurists and generals of Rome, to the merchants of Victorian England and the industrial entrepreneurs of America, these groups have waxed and waned, leaving their imprints on history. They have affected not only the political, social, and economic behavior of nations but also their spiritual and ideological attitudes. Marx tried to straighten the complex web of history by assigning to each society a dominant economic class and declaring religion, philosophy, and art servants of the economic interests of this master class. One does not need to accept this crude simplification to recognize the importance of the material interests and spiritual attitudes of certain—not necessarily economic—population groups in different historical periods.

Our society is more complex and eclectic. Many different groups influence American public attitudes and affect the paths of American statecraft. Their group influence is superimposed on the egalitarian and anonymous mechanism of formal democracy—one man, one (secret) vote. The practical idealism of the organized churches, the constructive pragmatism of industrial entrepreneurs, the welfare ideals of the labor unions, the law-centered attitudes of the legal profession, are among these influences. Until recently, however, one voice has been almost mute in this chorus—the voice of science. The influence of scientists, with their particular brand of sceptical idealism and their exploratory approach to the problems at hand, has been weak, if not absent, from the councils of free society.

In his televised farewell address, President Eisenhower warned the American people against certain new forces in American public affairs. He named as such the professional military and the arma-

Paper contributed to the Massachusetts Institute of Technology Centennial Conference on Science and Engineering Education, March, 1961.

ment-makers. A powerful peacetime military establishment and the great armament industry supporting it are, he warned, out of keeping with American tradition, even though they have been made unavoidable by the new development of military technology. Equally contrary to American tradition, he continued, is strong governmental support of science: it, too, has been imposed on us by new exigencies of international struggle. These newly created forces must not be permitted to exercise undue influence on the affairs of state. The religious, humanitarian, and civilian foundations of American society must be preserved. It did not occur to President Eisenhower to welcome science as a powerful ally in the contest for the satisfaction of human needs and allegiance of human minds now being waged between the forces of political and spiritual freedom and those of dogmatic regimentation. Even less did it occur to him to look to science for the resolution of this conflict.

Seeing in science a threat to the traditional structure of Western libertarian society is not restricted to America. As C. P. Snow said in his much debated 1959 Rede Lectures on "Two Cultures," classic education, based on the study of philosophy, history, and law, remains the only accepted foundation for the formation of a British statesman. Knowledge of the fundamentals of science and technology is not only neglected but considered unworthy of a prospective political leader and appropriate only for technical experts, who—to use the expression attributed to a British statesman—should be "on tap and not on top." Yet, the challenges now facing mankind and the tasks confronting our statecraft are largely the handicraft of science and technology, and scientific understanding and competent technical judgment are needed to meet them. It is not enough to be assured that in situations of this kind scientists and engineers will be called into consultation. If scientists had waited for the military to consult them as to whether "an atom bomb could be made on the basis of the discoveries of Hahn and Strassman," America might not have had nuclear weapons to this day. And although one is tempted to say that this would have been a blessing, nations not letting their scientists speak up would soon find themselves in grave jeopardy from those who listen to theirs.

When in 1948 the senator from Iowa accused the Atomic Energy Commission of "gross mismanagement," he refused even to ask whether the Commission was doing well in the development of nuclear energy for military and industrial purposes. He did not understand such matters and saw no reason why he should try to understand them; what he was concerned with (and felt himself competent to judge) was the excess cost of garbage cans in Oak

Ridge and insufficient security measures at Argonne. Legislators and statesmen of this kind, who not only lack technical judgment but consider it none of their business to try to acquire it, are not qualified to make many of the vital national decisions in our time. In military and strategic planning, in the estimation and management of natural resources, in the determination of the technological possibilities of assistance to new nations, and in many other fields, informed decisions can now be made only by those who (with the help of experts) can properly appreciate the scientific aspects and potentialities of a situation.

Those for whom science is a mysterious black box from which answers can be obtained by dropping in nickels (or million dollar checks) are disappointed when these answers come out in the form of cautiously qualified probability estimates which do not eliminate the need for understanding interpretation and balanced political judgment. Under the pressure of such incompetent questioners, scientists may be induced to make categorical "yes" or "no," "black" or "white," statements, thus going beyond the tentative conclusions justified by strictly scientific considerations. If, then, one scientist says "yes" and the other "no," the questioners turn away in disgust and say that scientists contradict each other and that their advice is therefore useless. We have seen this happening recently in the question of the biological hazards of nuclear test explosions.

SCIENTISTS AS NEGOTIATORS

Despite the difficulties of communication and the widespread distrust of scientists in public life (compounded of a fear of science as black magic and belief in the scientist's utter impracticality), the pressure of major developments unleashed by science in our time has now begun to destroy the traditional separation of science from politics. The area of public life in which this breakthrough first occurred was, naturally, that opened by the invention of nuclear weapons—the arms race and its effects on our foreign policy. The scientists involved in the development of the atom bomb and deeply concerned with the danger it has posed for mankind—the "band of frightened men," to use Harold Urey's expression—have greatly influenced the gradual evolution of American foreign policy from smug reliance first on nuclear weapons monopoly, and then on "massive retaliation," to a search for arms control and disarmament. In the same way, the influence of Russian scientists has tipped the balance in the Kremlin against those who confidently believed that the next war would destroy capitalism and make communism victorious everywhere in favor of those who argued that a nuclear war would destroy capitalist and communist civilization alike. This example shows the unpremeditated parallelism in the influence of scientists on political decisions on both sides of the Iron Curtain—a parallelism made inevitable by shared scientific knowledge and common respect for facts.

Beyond this spontaneously converging influence on public opinion and government policies of the West and the East, scientists from both groups have been increasingly involved in official and unofficial conversations about the problems created by the release of nuclear energy. During the prolonged negotiations on atomic energy control in the United Nations Atomic Energy Committee from 1946 to 1948, only one document was unanimously agreed upon, the "Feasibility Report" prepared by the scientific experts of both sides. Ten years later, the Geneva Committee of Scientific Experts gave another demonstration of the capacity of scientists to find a common language, despite allegiance to different ideological camps, by producing a unanimous report on the requirements for objective verification of a nuclear test ban. Admittedly, this unanimity was broken a year later by the American discovery of the possibility of muffling underground explosions. However, even in this case, trouble arose not from a disagreement among scientists about the technical aspects of the problems but from political differences in the estimation of the importance of muffling for the acceptability of a test ban agreement. Although scientists could not bridge this disagreement, they could and did agree on the need for further research—preferably internationally sponsored research—to establish the exact limits of the detectability of underground explosions with available instruments and to improve their sensitivity.

Prime Minister Macmillan was so impressed by the agreement reached by the Geneva Conference of Scientific Experts that in a speech to the United Nations in October, 1960, he proposed the creation of an international body of scientists to study all future disarmament proposals before they are submitted to diplomatic negotiations. There is much to be said for this approach, but no miracles can be expected from it. As with nuclear weapons tests, the critical problems of controlled disarmament are not technical. They are largely political; and if scientists could contribute something toward solving problems of this type, they could not do so within the confines of purely technical studies. All that such studies can achieve is to clear away the underbrush of technical difficulties obscuring the ultimate political problems. Scientists can take the technological uncertainties out of the political controversy; and they can work together on improving the technological situation, not only because they operate on the basis of mutually accepted facts and by mutually accepted scientific methods of reasoning and experimentation, but also because they have a common motivation —to enlarge the area of scientific knowledge and to improve the application of this knowledge to the technical problems at hand.

Many would say that this is where the function of scientists in negotiations on disarmament and arms control ends, and statesmen and diplomats must take over. Many scientists, however, would argue that scientific considerations, although intertwined with those of politics and strategy, continue to be significant far beyond the limits of purely technical inspection problems. In the estimation of the acceptability of a test ban agreement without an infallible inspection system—and no infallible control system is in sight— the questions of political trust between nations, on which scientists have no specially valid judgment, are coupled with the estimation of the advantages which each side could derive from clandestine tests. Scientists cannot provide an exact evaluation of these dangers, but their guess in this field may be more objective and informed than that of professional diplomats, or even of military experts, because the latter traditionally operate only with weapons already on hand or on the drawing board, whereas scientists can use a broader scientific imagination—the imagination that has led them to foresee the possibility of the fission and the fusion bombs.

The matter goes even deeper. Not only the weapons developments but also the likely strategies of the major powers must be evaluated in order to decide on the acceptability of a test ban; and in this field, too, scientific imagination can make important contributions, ranging beyond the historically conditioned imagination of military planners. Many ideas which are now fermenting in American strategic planning councils have had their origin in the imaginative thinking of scientists.

Many Americans scientists are now of the opinion that, in the face of the dangers of a nuclear war, the most advantageous strategy for both the United States and the Soviet Union may involve the maintenance of a limited, invulnerable deterrent capability and the renunciation of the development of an antiweapon capability and —according to some—even of attempts to protect the civilian population, and that the way to make this strategy work is to provide a broad exchange of information with our potential enemies on the chosen strategy and available arms, renouncing the traditional advantages of secret weapons development and deployment. Although less easily than the technical problems of weapons controls, these strategic matters can be discussed with some degree of objectivity and mutual understanding between scientists belonging to opposing sides. This is, in fact, the hope of some American scientists who seek opportunities for private discussion with their Soviet counterparts—such as those provided by the COSWA[1] (Pugwash) meetings. They would like to discuss not only the technical problems of disarmament but also the policies and strategies which could minimize the consequences of an evasion of arms control agreements and reduce the possibilities of accidental breakdown of a security system.

However helpful such discussions and investigations of weapons-control and military-strategy problems may be—and their alternatives are either an unrestricted armaments race or disarmament agreements concluded by diplomatic bargaining without scientific evaluation of their consequences—their potentialities should not be overestimated. Scientists can make their conclusions only by assuming that each side will act rationally to maximize the probability of achieving its aim. At best, they can correct this postulate of rationality by assuming a certain range of uncertainty due to irrational actions or random unforeseeable events. They cannot take into consideration the basic irrationality of men making political decisions or the impossibility of isolating one aim toward which a nation is supposed to strive—such as security from war—from other aims—such as preservation or extension of national or ideological power (not to speak of the personal ambition of a political leader)—which may loom more importantly on critical occasions.

Need for a New Reformation

It is a commonplace that the scientific progress of mankind has run away from the advancement of its traditional social organization and from its atavistic psychological attitudes. Under these conditions of "cultural lag," hoping to achieve permanent peace through disarmament amounts to "saving mankind despite itself." Many arms-control and world-security plans, invented by ingenious minds (many of them scientists), look almost like "gimmicks" intended to prevent nations from killing each other off without giving up their urge to kill.

Because the situation is terribly dangerous, attempts to find such shortcuts toward a peaceful world must be pressed forward with all possible diligence, and even gimmicks, however paradoxical, need to be explored; but one should have no illusions about the probability of reaching stable solutions on this precarious path. The

[1] Conferences on Science and World Affairs.

prevalence of irrational or pseudorational attitudes in international relations creates a framework within which rational solutions of single problems—even if they are as critical as that of the nuclear arms race—are well-nigh impossible. In the long run it is the framework itself that needs to be changed. What is needed is a change in morality which would make mankind fit to live in the nuclear age! Ultimately, this reformation will have to be crowned by the creation of bodies that institutionalize permanent peace, but a change of heart must precede a change of institutions.

It has been suggested that the only possible way toward a new morality in international relations would be through the growth of religious conscience, which would bring to practical realization the now largely unfulfilled tenets of the great religions—the universal brotherhood of man, the sinfulness of all taking of human life, the subordination of selfish impulses (of an individual or of a nation) to the common interest of mankind. Whatever one may feel about the need for the emergence of such a new religious spirit, hoping for it does not obviate the need for rational action directed at the same aim—a change of human attitudes without which the existence of man in the nuclear age will remain precarious. One could even argue that, by some kind of relation which physicists may be tempted to call "complementarity," the spiritual progress of mankind occurs in parallel with the development of rational answers to the physical problems of human survival; but this is not the place to argue this matter.

The main point I want to make in this paper is that science can make significant contributions to the needed re-education of mankind. Its first successful contribution was to enlighten nations and their rulers about the dangers of scientific war. Scientists have not yet succeeded equally well in making nations aware that merely junking scientific weapons, or even all weapons, will not provide security against a nuclear war, since the knowledge of how to make these weapons—and, therefore, the capacity to produce them in the emergency of war—will remain with mankind. Many scientists, not only in the Soviet Union but also in America, still believe in the simple formula: no nuclear weapons, no nuclear war; no weapons of any kind, no war at all. However, the recognition is growing that what needs to be abolished are not only certain weapons, or all weapons, but the institution of war itself. Scientists from all countries, East and West, have proclaimed this in the Vienna Declaration (issued after the Pugwash conference in Kitzbühel in 1960); and this recognition is spreading in the public opinion of most nations.

While instilling in mankind a holy fear of war is a good thing, it is only a first step. The next step must be the recognition that war has become not only too dangerous but also unnecessary in our time, because the reasons which have justified wars in the past have become invalid now. Man cannot survive by merely being frustrated in his desire to make war; he must become accustomed to thinking of his future in a framework which excludes war; and science offers him new alternatives.

Two motives have caused wars in the past. One was based on the experience of mankind that the earth offers only a limited amount of wealth to support the life and prosperity of a group of men—tribe, nation, or state. To permit its growth in numbers or in prosperity—and often simply to permit the maintenance of existing numbers and the already acquired standard of living—one group has had to take wealth away from other groups, whether in land, slaves, or natural resources. In our lifetime, the German aggression against the East was justified by the need for land on which to settle German farmers, and the Japanese aggression against the continent of Asia by the pressure of the Japanese population. The Anglo-French attack on Suez was justified by the necessity to assure the flow of Middle Eastern oil to the nations of Western Europe that were becoming deficient in fossil fuel.

However, when the military attack on Suez was stopped by the combined effect of world disapproval and the Soviet threat of nuclear war, the British government ordered a tripling of its nuclear energy production program. Nuclear power is not a complete substitute for power derivable from oil; nevertheless, this reaction is symbolic of the new age. Instead of trying to assure by force mastery over the limited reserves of natural wealth, nations must now attempt to produce new wealth by the application of science and technology. In the same era in which science has made war a potential suicide, it has offered mankind an alternative—creation of wealth, instead of its conquest. This change in creative capacity is less sudden and spectacular than the change in the destructiveness of weapons available to man. For some time to come, nations will remain dependent for their prosperity on assets such as fertile land and an abundant supply of iron ore and fossil fuel, but the transition has begun. With every year, it will become clearer and clearer that the future belongs not to nations in possession of the most land or the greatest natural wealth but to nations capable of the greatest scientific and technological effort.

The new importance of scientific and technological potential and of the capacity for its application must also affect the traditional concepts of the importance of certain economic structures and production systems for public welfare. In the nineteenth century, it may have seemed that human productivity was so limited that if too much of the produced wealth was retained by the owners of the tools of production the workman was bound to be left with next to nothing. Since then, technological progress has increased productivity to such an extent that it now has to be, by necessity, distributed among the widest groups of population; the upper classes alone simply cannot drive all the automobiles, live in all the houses, wear all the clothing, which American industry can produce. Even the whole population of America could not consume all the grain, meat, and cotton that our agriculture can produce. The pressure of high productivity has caused the state to become concerned with the maintenance and growth of the buying capacity of its population. In a continuation of the same trend, the only adequate solution of the present crisis arising from our excess productive capacity is to produce for the untold millions of people whose basic needs now remain unfulfilled. Thus (as recently suggested by the British Labourite A. R. Crosland), the welfare state of today is developing into the welfare world of tomorrow. Communists refuse to accept these facts, which contradict the predictions of Marxism; but this pretended blindness cannot continue indefinitely. The greater the influence of scientists on Soviet attitudes and policies, the more respect these policies will have for facts and the less heed will be paid to dogma. The latter will not be abandoned—the theory will be elaborated to explain away the obvious discrepancies between fact and prediction—but actual policies will be increasingly divorced from these ritual obeisances, and the stability and capacity for technological progress of modern capitalist economies will be tacitly taken for granted.

Similarly, the widespread American belief that only a competitive economy, based on private initiative and the profit motive, can achieve high productivity needs revision. The efficiency of nationally owned railroads in European countries and the economic prosperity of the semisocialist Scandinavian countries could, perhaps, be dismissed as minor phenomena; but the industrial upsurge of the Soviet Union cannot be similarly disregarded. Industry not spurred by the profit motive and competition may not be able to operate quite as economically or be as sensitive to the needs of the consumer (be it an individual or the state) as industry put on constant alert by these motivations (at least if truly competitive, which is not always the case, as proved by the recently revealed price collusion in the electrical industry!). More important, however, is the demonstration that industry deprived of profit motivation can nevertheless develop to high technological accomplishments. In some fields, such as the development of scientific weapons or the building of rocket ships, state leadership may be even more efficient than a competitive private effort. (The opposite seems to be true in agriculture, where attempts to eliminate individual profit motivation have so far been utterly unsuccessful.) What matters most from the point of view of achieving an abundant supply of goods is not the dogmatic differences between the two forms of production, but whether, under given special conditions, the one or the other is better able to make full and prompt use of the scientific and technological possibilities of the situation. (Another question, the answer to which is anything but simple, is whether there is a close relation between freedom of economic competition and the political and spiritual freedom of the individual, because the latter freedom is in itself a value as important as economic abundance—and becomes even more important once the minimum economic needs of the people have been met.)

The second reason for wars in past history has been the belief, carried by many segments of mankind, that its faith or ideology was the only correct one, that its economic or political organization was a supreme achievement, and that the good of its race, or creed, or state, was the supreme good. Wars have been fought, without any economic motivation, simply to spread a religion or a political system at the expense of other, supposedly inferior ones; and wars to increase the wealth of a nation received additional justification by serving also to spread a religion, or the power of a superior race, or the sway of a political or economic system.

In this field, too, the influence of scientific culture on human affairs can be salutary. The internationalism of science is traditional. Scientists belonging to different nations or different religious or ideological camps are not enemies or competitors but collaborators in a common quest for the understanding of nature and the utilization of this knowledge for the benefit of man. This is due not to a moral superiority of scientists but to the nature of their activity. They have the same aims, use the same approach, and depend for their success on exchange of experience with all others working in the same field. The value of world-wide co-operation and the threat of stagnation implied in any attempt at scientific secrecy and isolationism are obvious to them. Modern science began when the secret procedures of alchemy were replaced by open communication of results and theories. Science has found in the complete openness of its proceedings, giving everybody the chance to correct the errors of everybody else, an effective guarantee of continuous progress.

It is through this open collaboration, reaching across all national or ideological borders, that science has become the only field of human endeavor which knows no ups and downs but only slower or faster advancement. Attempts to declare the science of a certain race, or of a certain ideological system, different and superior to others, have perverted some scientists, who have proclaimed Aryan science superior to Semitic science or asserted that Communist science is in possession of an infallible guide to truth—dialectic materialism—which prevents it from making the errors into which the "idealist" science of the capitalist world is bound to fall. However, these outside perversions have never made—and cannot now make—a lasting impression on the scientific communities of the world, every member of which is continuously exposed to a contrary experience. This is what makes science one of the strongest forces in existence that work toward the integration of mankind into a single community through co-operation in the achievement of common aims. Science does not belong to any one country or any one political or economic system; it is the only common enterprise of mankind.

Congenital internationalism is not the only aspect of the scientific culture which makes its influence important for the re-education of mankind critically needed to assure survival in the nuclear age. Another and perhaps even more important aspect is the attitude of humility in the face of the unknown—of uncertainty as to the ultimate truth—which is the fundamental attitude of science in our time. This assertion may cause some raised eyebrows. Scientists are often considered as arrogant rather than humble, believing in their capacity to solve all human problems by their supposedly infallible methods of scientific reasoning and experimentation. This may have been true in the nineteenth century, when some scientists, at least (and many non-scientists impressed by the progress of science), did believe that classical mechanics could, in principle, explain all human experience. We see in our time the vestiges of this arrogance in the allegedly "scientific" philosophy of dialectic materialism, with its claims of possessing a clue to all physical, biological, and social phenomena. But science has long outlived this early and immature arrogance, replacing it by humble recognition of the limitations of our present knowledge of the universe and the relative and temporary character of all generalizations made on the basis of this knowledge. From the immensity of the universe and the tinyness of our own place in it to the complexity of the unimaginably small world of elementary particles, a modern scientist is continuously made aware of how little he as yet knows, even though he is confident of the capacity of science to learn much more in the future. He is mentally prepared to see many of the fundamental tenets and concepts he has assimilated in school found wanting as our knowledge of the world unfolds—as the law of the conservation of parity was proved wanting a few years ago. A scientist must always be prepared to submit his beliefs, findings, and generalizations to the never ending test of observation and experiment. Not that he is entirely without resistance to new theories that would overthrow the principles which he has become accustomed to accepting as valid; but of all groups of men, he belongs to the most open-minded one, the one most ready to accept change. He would be a poor scientist who would refuse to consider new facts and to change ideas to accommodate them. The only thing of which science is intolerant is intolerance itself—claims that certain concepts are sacrosanct, true beyond doubt, and protected from the test of logic and experience. Science is therefore a powerful antidote against the stubborn dogmatism which poisons the relations between different segments of humanity. The greater its influence on human affairs, the better the chance of bridging these clefts.

The belief of scientists in the practically unlimited capacity of scientific thinking and experimentation to increase man's understanding and mastery of the material world does not imply a denial that the world has aspects which the scientific approach cannot encompass, and that other paths can reveal them; but the more science is aware of the limited, relative, and temporary character of its own knowledge about nature, the less it is inclined to accept the concepts derived from other kinds of human experience as absolute and final rather than also as approximations to truth. Under the influence of scientific scepticism, moral values and judgments do not disappear; the concepts of beauty and goodness, of the spiritual essence of human existence, and of belonging to some spiritual order of things are not destroyed; but the claim of any individual or group to possess the ultimate spiritual truth, in full clarity and detail, appears as arrogant as would be the claim of a scientist to possess complete understanding of the physical aspects of the universe. Admittedly, the teaching of science in our time is not consciously directed at the fostering of this tolerant, evolutionary and relativistic attitude, but it could and should be so directed.

Not all scientists are able to extend the mental attitudes on which their achievement as scientists is based to other fields of human endeavor; but scientists as a group are more likely than any other group to approach human affairs with a considerable degree of open-mindedness. In the discussions of the problems of arms control and disarmament in the recent Pugwash meeting in Moscow, the most exhilarating moments occurred when such an open-minded approach was brought to bear on some of the controversial questions. These moments were rare, but they are what makes such discussions worth while.

Rereading what I have written, I note that it could make the impression of an uncritical panegyric to science as a power to solve all human problems, including those of distrust, conflict, and war between nations, and to scientists as the men most able to advance this aim by negotiations. This, most emphatically, was not intended. Scientists as human beings are as remote from being ideal incarnations of the sprit of science as, say, lawyers are from incarnating the spirit of law or ministers the spirit of religion; and the scientific method itself is as vague a concept as, say, the legal or the religious approach to human problems. This does not mean that these concepts do not correspond to something real and unique in the spiritual achievement of man, and do not represent very real forces influencing human attitudes and actions, or that the professional groups trained in these fields—the scientists, the lawyers, the clergy—do not possess, on the average and granted the widest possible exceptions, some understanding of the peculiar spirit of their profession and greater commitment to its application to human problems. All that I am pleading for in this paper, as pointed out at its beginning, is the need to give the spirit of science a stronger bearing on the formation of national attitudes and the making of public decisions than it has had in the past.

Heroes of Our Time

The world has a new set of heroes whose fame extends far beyond their native countries. They are called cosmonauts in Russia and astronauts in America—Gagarin, Titov, Glenn. (In justice, Shepard and Grissom should be added to this list, since their flights into the stratosphere, however brief, included the most dangerous moments of orbital flights, the launching and the descent back to earth.) The three who have acquired fame—Gagarin, Titov, and Glenn—are strikingly similar in character and background. They are bright, intelligent, well-bred, and well-educated men, self-disciplined, with a clear purpose in life. They are good family men, with attractive wives and children. James Reston, in the *New York Times,* pointed out the contrast between these new heroic examples, to be followed by American youth, and the restless, neurotic heroes of television and film. John Glenn, and probably other American astronauts as well, are faithful churchgoers; Gagarin and Titov are good Communists. Far be it from me to equate religion with the secular creed of communism, but both connote conformism to the approved national attitude. Not so long ago, in a Gallup poll, Americans were asked whether they would consider a Negro, a Jew, a socialist, a communist, or an atheist as a possible President of the United States. The atheist fared worst of all, despite the fact that one of the most illustrious American Presidents, Jefferson, was very close to being one. Churchgoing is in today's America a criterion of good citizenship, of unquestioned acceptance of certain general principles of human existence which have become associated, in the minds of a vast majority of Americans, with the American way of life. The Supreme Court may have quashed the Maryland statutes which barred atheists from occupying any state position, but in our time the reference to God has been introduced into the pledge of allegiance to the flag without much opposition. At the same time, in the Soviet Union, being a Communist has ceased to be a fighting creed and has become a matter of good citizenship as defined by the Communist state—leadership in youth organizations, volunteering for community activities, good scholarship, and uncritical commitment to state policy.

The Gagarins, the Titovs, the Glenns, well deserve their status of

From *Bulletin of the Atomic Scientists,* May, 1962.

world heroes; but in their achievements they were servants of their states and of the societies which produced them. Gagarin and Titov have become the best Soviet ambassadors at large, living advertisements for the advantages of the Communist system and the wisdom of Communist leadership. Glenn is bound to become a similar ambassador of good will, representing to the world the achievements of free American society.

None of the three heroes is an originator or creator. They are typical of a whole generation of heroes of our time—drivers of race cars, fliers of fighter planes, test pilots. They are daring handlers of machines designed by the genius of others. Typically, the names of the builders remain unknown. They are large teams in which no one is singled out in the public mind. The builders of Friendship VII are almost as anonymous as the builders of Vostok I. The name of the German constructor, Wernher von Braun, was singled out for a short time in the American public mind but has already been submerged in the large family of designers and engineers working for the United States space program. And truly, are they the creators of the space ships? Or are the creators the theoreticians, the professors of aerodynamics and ballistics, whose names are hardly known to the public? Mstislav Keldysh, the new president of the Soviet Academy of Sciences, may be one of them, but we were never even told whether his work was directly related to the development of the first Sputnik or its successors.

The clean-cut, sympathetic family men, the astronauts and the cosmonauts, resemble each other closely, whether they come from America or from Russia. As other parts of the world enter the adventure of space exploration, they are likely to produce similar men, who have been found most likely to have the steadiness, the ability to act wisely, and even the capacity to relax under the trying conditions of the adventure in space. Left to themselves, it would be easy to imagine Titov as a member of the American group of astronauts or Glenn as one of the Soviet cosmonauts.

These men could probably qualify as members of a new worldwide brotherhood. I listened in Moscow to an interview with Titov soon after his flight, and later in America to that with Glenn after his return from space, and the similarity between the two men was striking. And yet these heroes serve, faithfully and intelligently, two political powers that face each other with deep enmity and hope for the collapse of each other. It is easy to imagine the same type of man carrying, if so ordered, megaton nuclear bombs to obliterate Moscow or to incinerate New York. These new world heroes of the age of science are obedient servants of prescientific national states.

I read once, I do not remember where, a novel describing the ultimate revolt of flyers of all countries, who refused to continue slaughtering innocent people from the air and formed an international government of the air age. But chauffeurs and pilots are even less likely to rebel against those who own the machines they are driving or flying than the scientists and engineers who discover the laws on which these machines are based and build them at the bidding of their governments.

Scientists, too, form a single fraternity; and engineers and technologists the world over are similar in their attitudes and outlooks, even if they are not so accustomed to world-wide co-operation as scientists. They are all potential useful citizens of a peaceful united world, as are the test pilots and the astronauts; but tradition keeps them all subservient to the power interests of the separate states. They serve these states in constructive pioneering work and exploration, as well as in the development and use of new weapons of man's destruction.

The scientific revolution of our time has made national states obsolete, whether they are based on race, tradition, creed, or ideology; but before mankind has realized this obsolescence, this revolution has provided the leaders of the different states with technical facilities sufficient to destroy their own societies and to drag all mankind down into obliteration with them. It has given man a source of energy sufficient to produce an adequate standard of living for all—or to incinerate the largest city in a single explosion. It has given man vehicles which could be used to explore the universe far beyond the dreams of any preceding generation—or to carry megaton bombs from one end of the earth to another with the precision of an archer hitting a target at fifty yards. It has developed men who are ready to do both. These are the heroes of our time—the Gagarins, Titovs, Glenns, and the many others following in their footsteps. Societies are as yet only dimly aware that they must use these men (and their machines) for a common effort on behalf of humanity as a whole and not for their internecine fight. The exchange of telegrams between Khrushchev and Kennedy after Glenn's flight, both agreeing to the desirability of international collaboration in space exploration, are good words; but are the two societies ready to follow them with actions? In 1961, I wrote in the *Bulletin of the Atomic Scientists* that a Russian and an American astronaut flying together toward some cosmic objective would do more for the mutual understanding of the two nations than years of diplomatic wrangling. This day seems only a little less far out of sight now than it was a year ago.

Index

Adams, Sherman, 30
Adenauer, Konrad, 179
Africa, 76, 88, 89, 142, 143, 174, 177, 213, 233
Antarctica, 50, 238
Antimissile missiles, 74, 165; strategy involving, 243; *see also* Missiles
Arms control, 59–60, 95
Arms race, 53–54, 60, 103–4, 227
Art, relation of, to science, 298–304
Asia, 76, 88, 142, 143, 173, 185, 213
Astronauts, 325–27
Aswan dam, 213
Atomic Energy Commission (AEC), 277–81, 313
Atomic power, 33, 109, 132–33, 247, 310, 319; international control of, 20, 104, 107–8, 115–16, 132–33, 296
Atomic weapons, 12–15, 19, 56, 99–111, 131–32, 165, 241, 244, 251, 253; *see also* Weapons
"Atoms for Peace," 137, 223, 255, 315

Bakh, A. N., 195
Baldwin, Stanley, 121
Baratynski, Evgeniy A., 299
Baruch, Bernard, 117, 118
Beria, Lavrenti, 85–86
Blok, Alexander, 48
Boas, George, 291–95
Bohr, Niels, 201, 226, 291
Borgese, G. A., 48
Born, Max, 201, 218
Bulganin, Nicolai A., 50, 142, 147, 150–51, 163

Canada, 102, 229
Cavers, David, 21, 133
Childs, Marquis, 163
China, 64, 66, 70, 75, 82, 88, 90–98, 173, 212–13, 312
Churchill, Winston, 73, 75, 138, 173
Civil defense, 135
Clark, Grenville, 48
Common interests, of mankind, 28, 40, 46, 64, 228–30, 236, 271, 307
Conferences on Science and World Affairs (COSWA-Pugwash), 44, 47, 50, 54, 60, 160, 176, 189–92, 193, 209–57, 319; in Baden, 44; in Kitzbühel, 217–26; in Lac Beauport, 209–16; in London, 245–50; in Moscow, 176, 227–44; Vienna Declaration of, 47, 50, 54, 60, 176, 191, 238–39, 249, 251–57, 319
Congo, 174, 232, 233
Conventional weapons; *see* Weapons, conventional
Co-operation, international; *see* International co-operation
COSWA; *see* Conferences on Science and World Affairs
Cuba, 57–69, 70, 76, 78, 83, 98, 173
Czechoslovakia, 85, 123, 150, 154

Daniel, Cuthbert, 133
De Gaulle, Charles, 39, 53, 84, 85, 97, 177, 231
Dedijer, Stevan, 34, 77
Deterrence, 34, 42, 139, 141, 146–59, 163–64
Dillon, C. Douglas, 148
Disarmament, 20, 21, 22, 41, 44–46, 60–62, 93–94, 97, 141, 156, 160–61, 178, 190–91, 227–44, 252, 316, 318
Dispersal, of cities and vital industries, 18, 103, 104, 135, 272
Dulles, John Foster, 150, 152, 178

Eastern Europe, 80, 85, 152, 177, 211, 228, 232
Eastern Germany, 63, 152, 182, 232
Economics, 221, 321
Education, 261–65, 282–97; *see also* Science, and education
Egypt, 147, 148, 149, 158, 213, 232
Einstein, Albert, 201, 204
Eisenhower, Dwight D., 21, 50, 129, 137, 141, 171, 172, 174, 177, 179, 269, 313, 314
Ethics, 25, 144, 220, 317–18, 323
Europe, in 1954, 120–30
European Defense Community (EDC), 25, 121

Fallout, 16
First- and second-strike weapons; *see* Weapons
Flanders, Ralph, 158
Ford, Henry, 77
France, 84, 120–21
Franck Report, 99–112, 131, 215
Franklin, Benjamin, 261
"Freezing" national boundaries, 65, 83, 124, 163, 210, 231

Gagarin, Yuri A., 325, 327
Gallant, Joseph, 291
Gardner, Trevor, 29
Genetics: controversy over, in Russia, 180–81, 199; and radiation dangers, 16–17, 254, 255
Geneva Conference; *see* "Atoms for Peace"
German scientists; *see* Scientists, German
Germany, 39, 65–66, 83, 85, 123, 179, 211, 231; *see also* Eastern Germany; Western Germany
Glenn, John H., 325–27
Goethe, Johann Wolfgang von, 96, 300
Golube (Soviet engineer), 238
Great Britain, 84, 120–21, 157, 221, 312
Grodzins, Morton, 209
Groves, Leslie R., 32, 116
Gruenther, Alfred M., 19, 147, 149, 150–51

Hearst, William Randolph, 161
Heisenberg, Werner, 201, 218
Hiroshima, and Nagasaki, 115, 118, 131, 146, 202, 295
Hitler, Adolf, 72, 182, 184, 219, 268, 283
Hughes, Don, 99
Humanities, versus science in education, 282, 297
Hungary, 65, 70, 85, 150, 152, 153–54, 155, 162
Huxley, Aldous, 291–92

India, 64, 70, 76, 102, 157
Indonesia, 92, 117, 143
Inglis, David R., 21, 164
Inspection, for disarmament, 141, 160, 228
International Atomic Energy Agency (IAEA), 50, 137, 176–77, 221, 256
International co-operation, 38, 40, 46, 47, 50, 52, 54, 66, 67, 191, 193–94, 236, 238, 244, 255–56
International Geophysical Year (IGY), 28, 46, 50, 161, 177, 237, 255
Iran, 88, 93, 174
Iraq, 88, 174, 232
Israel, 65, 312
Italy, 84, 213

Japan, 131, 182
Jungk, Robert, 218

Kahn, Herman, 73
Kennan, George, 163, 210
Kennedy, John F., 58, 171–79, 233, 327
Khrushchev, Nikita S., 38, 41, 44, 50, 52, 55, 57, 78, 90, 161, 162, 165, 170, 202, 204, 210, 227, 228, 233, 327
Korea, 12–15, 26, 70, 91, 143, 183, 212, 273

Lamartine, Alphonse Marie Louis de, 299
Langmuir, Irving, 137
Laos, 64, 88, 143, 161, 174
Latin America, 76, 78, 90, 177
Lebanon, 174, 232
Leghorn, Richard S., 59, 161, 164, 209
LeMay, Curtis, 141
Lilienthal, David E., 116, 118, 133
Lippmann, Walter, 165
Lysenko, Trofim, 199–201

MacArthur, Douglas, 278
Macmillan, Harold, 50, 315
McNamara, Robert S., 58
Marx, Karl, 54, 76, 90, 197, 312
Middle East, 120, 143, 153, 213, 233
Missiles, 28, 56–58, 70, 139, 165, 270; *see also* Antimissile missiles
Morgenstern, Christian, 180
Moscow, 143, 150, 155, 182, 195, 326
Muller, Hermann J., 164, 198

National interests, 52, 153, 229, 305
National Science Foundation, 267–69
National sovereignty, 29, 37, 144, 153–54, 156
Nations, new, 39, 52–53, 77, 169–70, 175, 221, 232, 233–34, 244, 256, 272, 301
Nesmeyanov (head of Soviet Academy of Sciences in 1958), 207
Niebuhr, Reinhold, 284
Nixon, Richard M., 50, 52, 129, 243
Nobel prizes, 34, 195, 205
Noel-Baker, Philip, 178
Norstad, Lauris, 151
Nuclear test ban; *see* Test ban, nuclear

Oak Ridge (Tenn.), 313–14
Oder-Neisse border; *see* Poland
Oppenheimer, J. Robert, 17, 127, 218, 277–81

Pakistan, 92, 93

Panama, 157–58
Pasternak, Boris, 81
Pauling, Linus, 165, 199, 296
Philipps, Thomas R., 137
Poland, 81, 85, 152, 154, 179, 273; and Oder-Neisse border, 65, 83, 179, 211, 231
Power, Thomas S., 42
Power politics, 61–62, 119
Pre-emptive and preventive war, 41–42, 44, 116; *see also* War
Price, Derek, 289
Pugwash; *see* Conferences on Science and World Affairs

Resources, world, 309–10
Reston, James, 325
Ridenour, Louis, 36–37
Rockets, 33, 42, 43, 59, 214
Russell, Bertrand, 184–85, 190, 215
Russell-Einstein appeal, 190
Russian science; *see* Science, Russian and Soviet

Satellites, 28, 205
Science, 97, 263, 264; and art, 298–304; as common enterprise of mankind, 66, 236, 246, 264; and economics, 54, 220; and education, 30, 205, 261–65, 282–97; and national affairs, 266, 278, 288, 314, 316; Russian and Soviet, 31, 95, 134, 137, 195–208; and world politics, 175–76, 193, 245, 246, 290, 303, 312–24
Scientific advisors, 278, 288, 314, 316
Scientific revolution, 48, 95, 182–83, 192, 245, 247, 249, 285, 305
Scientists: attitudes of, 189, 235–36, 265, 292, 321, 323, 327; German, 101–7, 218, 296; in politics, 35, 193–94, 219, 224, 261–62; responsibilities of, 207–26, 257; Russian, 189, 191, 315
Seaborg, Glenn T., 99
Secrecy, 44, 101, 109, 135, 241, 242, 271, 321
Security risks, 277–81
Semenov, Nicolai N., 35, 189, 204
Sherwin, C. W., 164 n.
Sohn, Louis, 48
Southeast Asia, 92, 120, 177
Sovereignty, national; *see* National sovereignty
Soviet Academy of Sciences, 206
Soviet science; *see* Science, Russian and Soviet
Soviet Union, 75, 78, 80, 95, 102; armaments and strategy of, 19, 31, 43, 56–57, 61–62, 91, 102, 104; domestic policy of, 81, 82, 91, 95; foreign policies of, 56–57, 63, 64, 66, 81, 82, 83, 89, 91, 93, 95, 117, 145–51, 157, 182
Speier, Hans, 135
Squires, A. H., 133
Stalin, Joseph, 72, 77, 80, 199, 273, 287, 298
Stearns, J. C., 99
Stevenson, Adlai E., 166, 177, 213
Stimson, Henry L., 99, 131
Stowe report on co-operation in science, 191
Strategic Air Command (SAC), 42, 147
Strauss, Lewis, 33, 137
Suez, conflict over, 49, 64, 148, 149, 158, 162, 163, 308, 310, 319
Szilard, Leo, 45, 61–62, 99, 131, 299

Taiwan (Formosa), 179, 213, 232
Teller, Edward, 165, 218
Test ban, nuclear, 21, 95, 156–57, 164–65, 166, 214, 228, 239, 240, 252, 254, 315
Titov, Gherman, 325–27
Topchiev, Alexander V., 207, 208
Truman, Harry S., 135, 158
Turkey, 88, 163, 164

Underdeveloped nations; *see* Nations, new
United Nations, 123, 148–49, 157, 169, 177, 234, 275, 306–7
United Nations Atomic Energy Commission (UNAEC), 20, 116–18, 123, 133, 262, 315
United States, foreign and military policy of, 61–69, 73, 83, 95, 97, 116, 118, 120–30, 153, 157, 171, 172, 174, 179, 211, 273
Urey, Harold C., 314
U.S.S.R.; *see* Soviet Union

Vavilov, Nicolai A., 199
Vavilov, Sergey I., 200
Vienna Declaration, 251–57; *see also* Conferences on Science and World Affairs
Vietnam, 92, 143, 185, 212
Vigny, Alfred de, 299
Vucinich, A., 206

Waddington, C. H., 209
War, 23, 51, 73, 249, 251, 253; causes of, 52, 72–74, 77, 227, 272–73, 319, 321; limited, 251, 254; threats of, 49, 306
Weapons: conventional, 12–15, 151–52; first- and second-strike, 44, 57–58, 59
West Berlin, 63, 70, 83, 147, 177, 179, 181, 182, 183
Western Europe, 83, 84, 123, 149, 151, 273, 319

Western Germany, 120
Wiesner, Jerome B., 209, 210
World community, 156, 167, 168
World War, Second, 5, 8, 24
Wright, Quincy, 133

Youth, in the Soviet Union and the United States, 285–86

Date Due